R e

Thomas P. Hughes

Rescuing Prometheus

Thomas P. Hughes is Emeritus Mellon
Professor of the History and Sociology of
Science at the University of Pennsylvania
and a Fellow of the American Academy of
Arts and Sciences. His books include
*American Genesis: A Century of Invention
and Technological Enthusiasm.* Currently a
Distinguished Visiting Professor at MIT,
he lives in Philadelphia.

Also by Thomas P. Hughes

American Genesis: A Century of Invention and Technological Enthusiasm

Elmer Sperry: Inventor and Engineer

Networks of Power: Electrification in Western Society, 1880–1930

Lewis Mumford: Public Intellectual (with Agatha Hughes)

Rescuing Prometheus

Rescuing Prometheus

Thomas P. Hughes

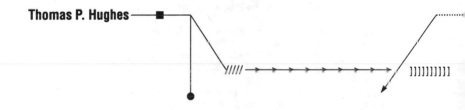

Vintage Books
A Division of Random House, Inc. • *New York*

Grateful acknowledgment is made to the following for permission to reprint previously published material: *Aviation Week & Space Technology*: Excerpts from "Human vs. Hardware" by Dean S. Warren (*Aviation Week & Space Technology*, June 7, 1971, p. 63). Reprinted by permission of *Aviation Week & Space Technology*. *The New York Times*: Excerpt from "Man in the News" (*The New York Times*, December 1957). Copyright © 1957 by the New York Times Company. Reprinted by permission of *The New York Times*.

The Library of Congress has catalogued the Pantheon edition as follows:
Hughes, Thomas Parke.
Rescuing Prometheus / Thomas P. Hughes.
cm.
Includes index.
ISBN 0-679-41151-8
Research, Industrial—United States. 2. Technology—United States—History—20th century. 3. Technology—Social aspects—United States. 4. Science and industry—United States. I. Title.
T176.H84   1998
303.48'3—dc21   97-44571
CIP

Vintage ISBN: 0-679-73938-6

Book design by Fritz Metsch

www.vintagebooks.com

Printed in the United States of America
10 9 8 7 6 5 4 3 2 1

For Agatha, whose spirit lives in this book

Contents

I   Introduction: *Technology's Nation*   3

II   MIT as System Builder: *SAGE*   15

III   Managing a Military-Industrial Complex: *Atlas*   69

IV   Spread of the Systems Approach   141

V   Coping with Complexity: *Central Artery/Tunnel*   197

VI   Networking: *ARPANET*   255

VII   Epilogue: *Presiding over Change*   301

Acknowledgments   307

Notes   311

Index   353

Rescuing Prometheus

> > > > > > > > > >

I Introduction:

Technology's Nation

Changes in the world's political, economic and technological realms in the past century have placed great stresses on approaches to management and system design. The changes that have had the greatest impact are the increase in size and complexity of the human organizations and technical systems needed in the world today, and the rate of change in the external environment with which these organizations and systems must cope.

Joel Moses, Provost, MIT

▼ During the late eighteenth and early nineteenth centuries, the United States was nature's nation. By the twentieth, it had become technology's nation. Americans had transformed a natural world into a human-built one characterized by technological systems and unmatched complexity. In doing so, they demonstrated a technological prowess unequaled elsewhere in the world. Although Americans take more pride in their political history than in their technological, the remainder of the world stands in awe of the technological transformations they have wrought.[1] The post–World War II period we shall consider witnessed the continuation of a technological transformation that can be seen as a second creation; the first was mythologized in the book of Genesis.

Telling the story of this ongoing creation since 1945 carries us into a human-built world far more complex than that populated earlier by heroic inventors such as Thomas Edison and by firms such as the Ford Motor Company. Post–World War II cultural history of technology and science introduces us to system builders and the military-industrial-university complex. Our focus will be on massive research and development projects rather than on the invention and development of individual machines, devices, and processes. In short, we shall be dealing with collective creative endeavors that have produced the communications, information, transportation, and defense systems that structure our world and shape the way we live our lives.[2]

This book tells of the design and development of four projects—SAGE, Atlas, Boston's Central Artery/Tunnel, and ARPANET—that exemplify the ways in which large systems are created, ones representative of post–World War II technology.[3] The SAGE (Semiautomatic Ground Environment) air defense project fostered the first interactive, digital computer designed primarily for information processing and process control rather than for computational functions. The SAGE Project exhibited a prime characteristic of post–World War II technology: engineers and scientists playing leading management roles as well as solving research and development problems. SAGE also exemplifies the collective management structure now labeled the military-industrial complex, or, more accurately in the case of SAGE, the military-industrial-university complex.[4]

The Atlas Project of the 1950s, which produced the first intercontinental ballistic missile, changed the complexion of both the Cold War and the aerospace industry.[5] To solve its extraordinarily complex problems, project engineers, scientists, and managers drew on a vast array of resources. During its most intensive design and development phase, Atlas involved 18,000 scientists, engineers, and technical experts in universities and industry; 70,000 people from office to factory floor in twenty-two industries, directly and actively participating; 17 associated contractors and 200 subcontractors with 200,000 suppliers; and about 500 military officers with technical expertise. These numbers not only suggest size but call attention to the heterogeneity of the project and the demand "to achieve a new degree of management coordination."[6] From Atlas, a mode of management known as systems engineering emerged and has spread throughout the military and industrial worlds and even into government agencies.

Boston's Central Artery/Tunnel Project (CA/T), begun in the late 1980s and continuing into the next century, has become the country's largest urban highway project. Like other large construction projects, it changes the built environment in which we lead our lives. In the course of the project, transdisciplinary teams solve engineering problems and manage the coordination and scheduling of the numerous subcontractors. By using a systems engineering mode of project management, engineers and managers of the joint venture firm of Bechtel/Parsons Brinckerhoff cope with even greater messy complexity than that present during the SAGE and Atlas projects. They must respond to countless environmental and other interest groups that help shape the design of the Boston project.

The ARPANET, a project of the Defense Department's Advanced Research Projects Agency (ARPA), opened the information highway. Initiated in the late 1960s, the project displays problem-solving approaches that have come to characterize recent projects, especially in the computer industry. These include a flat, collegial, meritocratic management style as contrasted with a vertical, hierarchical one; the resort to transdisciplinary teams of engineers, scientists, and managers in contrast to reliance on discipline-bound experts; the combining of diverse, or heterogeneous, physical components in a networked system instead of standardized, interchangeable ones in an assembly line; and a commitment by industry to change-generating projects rather than to long-lived processes. Like the other postwar projects, ARPANET depended upon a joint venture that combined government, industry, and university resources to provide the creative technological and managerial resources needed.

The reader may be surprised to find this history of technological projects focusing so often upon management rather than upon the engineering and science being managed. The author has chosen this emphasis because he believes that the engineers and scientists managing the projects have often found that management has presented more difficult challenges than research and development. Furthermore, he is of the opinion that the present-day preeminence of the United States in the creation of large systems arises in large part because of its managerial prowess.

]]]]]] Projects in History

This history deals with projects because they have been and are the organizational form best suited for the creation of technology.[7] The long history of projects extends back at least as far as the building of the great Egyptian pyramids and the Middle Eastern irrigation systems. The building of the Gothic cathedrals remains among the best-known projects of European history. Historians of art, of architecture, and of technology take these magnificent structures as their subject matter, for both aesthetic and technical considerations shaped the design and construction of these projects. Today the designers of the Central Artery/Tunnel Project face a similar challenge.

During the British and American industrial revolutions of the early nineteenth century, canals and railroads embodied the creative urge to design and order the environment on a large scale. Later, Thomas Edison's research and design project that culminated in the supplying of electricity to downtown Manhattan attracted public notice and celebration. In the early twentieth century, the construction of the Panama Canal, Henry Ford's design and development of the River Rouge automobile production plant, the Hoover Dam, and the Tennessee Valley Authority complex stand out as notable American projects.[8] Future historians will compare the projects considered in this book with those of Edison, Ford, and the other historical ventures, even though the names and creators of SAGE, Atlas, ARPANET, and the Central Artery/Tunnel are not yet widely known.

More recently, the World War II Manhattan Project, which produced the atom bombs, and the MIT Radiation Laboratory, which developed radar systems, have become exemplars of large-scale creativity. The system builders of postwar projects frequently acknowledge their indebtedness to those scientists and engineers who, in presiding over the Manhattan Project and wartime Rad Lab, showed them how to cultivate a curiosity-driven, imaginative, inventive style of doing research and development.[9]

System builders presiding over prewar projects assumed that

they would create a production, transportation, communication, or energy system that would remain virtually unchanged for decades; those presiding over technological projects today expect the systems they build to evolve continuously and to require new projects to sustain the evolution. This attitude is characteristic especially of engineers and scientists in the computer and communications industry. In a front-page article on 19 August 1996, the *Wall Street Journal* finally recognized the trend. Alluding to the increasing number of nomadic computer consultants who, moving from one project to the next, solve computer problems for various organizations, the article stated that the nomads point the way "toward something significant: More and more of the work in America is project oriented. . . ." Recent books about management focus on projects, too, especially those that create new computer hardware and software.[10]

System builders whom we shall encounter differ from the heroic inventors and managers of an earlier era. System builders preside over technological projects from concept and preliminary design through research, development, and deployment. In order to preside over projects, system builders need to cross disciplinary and functional boundaries—for example, to become involved in funding and political stage-setting. Instead of focusing upon individual artifacts, system builders direct their attention to the interfaces, the interconnections, among system components. Further, system builders often preside over the establishment of systems that involve both physical artifacts and organizations. In the history presented in this volume, system builders such as General Bernard Schriever of the Atlas Project and Frederick Salvucci of the Central Artery/Tunnel Project play leading roles in designing systems that incorporate both organizational and technical components.

Postwar system builders differ substantially from their pre–World War II predecessors in the size of the organizations over which they preside and in the collective nature of their activity. Schriever, Salvucci, Jay Forrester, Simon Ramo, and Joseph Carl Robnett Licklider, system builders whom we shall encounter, are the leaders of teams and organizations; they are deeply embedded in organizational structures. Timothy Lenoir points out that innovation, whether through projects or other technical and scientific endeavors, is essentially a distributed social process. Postwar system builders often function within the context of organizations dedicated primarily to system building such as the MITRE Corporation, the Western Development Division of the Air Force Research and Development Command, the Ramo-Wooldridge Corporation, the Advanced Research Projects Agency, and Bechtel/Parsons Brinckerhoff. The National Aeronautics and Space Administration (NASA) also performs as a major system builder and project manager.[11]

]]]]]] The Systems Approach: Managing Complexity

System builders presiding over SAGE and Atlas used an empirical, or cut-and-try, approach for solving managerial and engineering problems. Subsequently, academic engineers and social scientists articulated, rationalized, and organized the field experience garnered during such projects, thus presuming to create a codified managerial and engineering science, which they called systems engineering. In the 1960s, its proponents produced widely read articles, textbooks, and courses about the systems approach to technical and managerial problem solving.

Academic practitioners of systems engineering stressed its putative scientific nature, its quantitative approach, and its reliance on generalized theoretical foundations. Additionally, they quickly adopted the computer to process information, especially in a quantitative form. They also argued that academically, or scientifically, trained persons in the fields of management and engineering had especial expertise for implementing the systems approach.

Because of this emphasis on science and method, the postwar spread of the systems approach resembles the spread of scientific management early in the twentieth century under the aegis of Frederick W. Taylor and his followers.[12] Taylorites, however, applied scientific management techniques to ongoing manufacturing operations rather than projects.

In the 1960s, advocates of the systems approach developed a package of techniques and theory that included not only systems engineering but also systems analysis and operations research. While systems engineering was developed to manage large projects, operations research was designed to analyze military operations in place, and systems analysis was developed to analyze the anticipated costs and benefits of alternative planned projects, especially military ones. Together, these techniques have generated a managerial revolution comparable to that brought about earlier by Taylor's scientific management.

The package, or bundle, of systems approaches helped American system builders to cope with the complexity of technological projects and operations after World War II. Such capability has become a hallmark of American management and engineering. Robert White, formerly head of the National Academy of Engineering, which includes the elite of the professional world among its members, observes that maintaining excellence in engineering activities in the United States "means doing what we do so well— design and management of increasingly complex technological systems." He then characterizes "systems integration . . . [as] the ultimate test of engineering management."[13] Another observer declares that "the development of nuclear submarines, ballistic missile systems, and air defense systems . . . [has] emboldened society to tackle complex problems of immense scale, not only physical ones but social ones."[14]

]]]]]] The Military-Industrial-University Complex

To a society that conceives of itself and its technological achievements as emerging from a civil, free-enterprise environment, it may seem paradoxical that the widespread systems approach should have emerged from military-funded projects. Yet American history contains many instances of the military nurturing new technologies and new managerial techniques. The spread of the American system of production during the nineteenth century provides an early and outstanding example. Use of precision and interchangeable parts for gun manufacture spread from the government armory at Springfield, Massachusetts, to a host of metal- and woodworking establishments first, later to the manufacturing sector in general.[15] Early in the twentieth century, military funding initiated the intensive development of radio and airplane technologies as well as the application of complex control devices, such as the automatic airplane pilot.[16] More recently, military funding contributed enormously to the launching of the computer and communications industry, the commercial aerospace industry, and the electronics industry that flourish today.

A historian attempting to persuade a post–Vietnam War generation that the military-industrial-university complex has presided over a highly creative period since World War II faces a difficult challenge. A generation influenced by the counterculture values of the 1960s has difficulty in seeing the military-industrial-university complex as other than politicized, bureaucratic, and indiscriminate in the use of public funds, and destructive of the physical environment.[17] Proponents of the free-enterprise system also have difficulty appreciating the role of government in promoting technological change. Many Americans simply cannot conceive of the military as deeply shaping their culture and their values.

If, however, the historian focuses upon the period from 1950 to 1970, when military-funded projects dominated the innovative technological and managerial scene, a different picture emerges. Unquestionably, the system builders took little heed of the degradation of the physical environment at the sites where their systems evolved, but they energetically and effectively countered the bureaucratic tendencies common to large projects and organizations; they held at bay political forces that would subvert technical rationality. Motivated by the conviction that they were responding to a national emergency, they single-mindedly and rationally dedicated the enormous funds at their disposal to providing national defense. Opportunistic use of military funds to sustain regional economic development and corporate welfare came later.

Dispassionate accounts and analyses of the military-industrial-university complex remain rare, especially of its creative planning, designing, research, and development. Historians have yet to write broad-ranging histories that describe failures as well as successes, confusions as well as clear vision, issues unresolved as well as problems solved, corruption as well as high-mindedness, and paths not taken as well as those pursued. Historians someday will see the similarities between the half-century history of the military-industrial-university complex since World War II and other memorable eras of human creativity.[18]

From about 1950 to 1970 national defense projects attracted the cream of the professional world. In the 1950s, few young and established engineers, scientists, and managers chose to avoid defense work because of moral scruples. Financial compensation and professional advancement alone could not have stimulated the enthusiastic commitment to problem solving exhibited by the thousands of engineers, scientists, and managers during the early phase of the Cold War. They believed that they were responding to a threatening challenge from the Soviet Union.[19]

Alain Enthoven, who played a major role in cultivating the systems approach both at the RAND (Research and Development) Corporation and at Robert McNamara's Defense Department, speaks for many of his contemporaries in management and engineering:

> In the 1950s and 1960s, the United States and its North Atlantic Treaty Organization (NATO) were locked in a mortal struggle with the Soviet Union and its allies. The Soviets had taken over Eastern Europe, linked up with communist allies in China, armed client states in the Middle East, supported revolutions in the Third World, threatened repeatedly to use military force to drive us out of Berlin, and then covertly stationed nuclear-armed missiles in Cuba within easy reach of many of our bases. They preached a doctrine of worldwide revolution. The Soviets and their East European satellites posed an enormous threat to our European allies.[20]

From 1949 to 1953, *New York Times* editorials also portrayed the Soviet Union as a menace to the freedom and security of the

United States and the rest of the world. Several went so far as to equate Russia with National Socialist Germany as an aggressor nation. In general, the editorials judged the United States buildup of a deliverable nuclear arsenal necessary. A 1953 editorial, "Facing the H-Bomb," expressed the drift of the newspaper's attitude:

> While international agreement must be the basic area of such efforts [to avoid atomic Armageddon], much more is needed, too, in the way of official candor in making the cruel facts known to our people. Needed, also, is official recognition of the much greater resources that must be devoted to defense of our shores, a basic means of deterring aggression.[21]

Managers and engineers also believed that on large projects they could choose and work on problems that would advance the particular science or engineering fields that interested them, military relevance notwithstanding. Advisory committees consisting of civilian engineers and scientists often determined military research and development policy. In these cases, the military did not shape academic research and development activities as much as academics shaped military policies. Academic entrepreneurs used the SAGE and Atlas as well as ARPANET projects as means of pursuing research and development goals formulated in academia to advance fields such as electronics and computer science.[22]

The commitment of these professionals, however, began to erode in the 1960s during the Vietnam War and with the rise of counterculture values. Their idealism and enthusiasm evaporated rapidly with the indiscriminate defense spending in the 1980s and the attenuation of the Cold War.[23] Skepticism spread among engineers and scientists about the military, or defense, justification for weapons projects.[24]

Simon Ramo, head of the Ramo-Wooldridge Corporation, which served as systems engineer for the Atlas Project, observed a shift in attitude in the middle 1960s. He believes that the Vietnam War led many scientists to become less responsive to invitations from the Department of Defense to advise on or undertake research.

When President Richard Nixon, angered by the failure of scientists to support the war in Vietnam, stopped awarding the National Science Medal, dismissed his science adviser, and disbanded the President's Science Advisory Committee, the disillusionment with defense work deepened among academics.[25]

]]]]]] Public Projects and Responsibility

Before disillusionment spread, enthusiastic advocates of the systems approach that had proved effective during the SAGE and Atlas projects tried using it during the 1960s to solve social problems plaguing society. To their surprise and disappointment, the approach proved too fine-honed, too focused upon technical and economic factors, to respond to the messy, politicized urban problems targeted by President Lyndon Johnson for his Great Society Program and War on Poverty.

The 1960s became a watershed decade in the history of large projects and the systems approach. Their management and engineering style changed. Conceived in the late 1960s and created by young engineers and scientists, the ARPANET project, for example, differs from pre-1960s SAGE and Atlas in its expression of counterculture values, especially a rejection of hierarchical management structures. The military funded the ARPANET, but computer scientists and engineers presiding over the project pushed military goals to the background, emphasizing the spread of computer utilization and the development of computer networks as ends in themselves.

The post-1960s Central Artery/Tunnel Project has much less clearly defined and more politicized objectives than SAGE or Atlas, whose system builders valued technical and economic considerations most highly and often to the exclusion of other factors. While the military managers of Atlas successfully resisted politicization of the project, CA/T system builders tolerated and even embraced it as they responded to the demands of ethnic, environmental, and other interest groups.[26] Because it includes rather than excludes political and social factors, CA/T is an open rather than a closed system.

Although the market provides the public a voice in shaping consumer technology, the Central Artery/Tunnel demonstrates that the public can resort to political means to shape public projects. CA/T history also confirms that women and minorities can be involved in system building, previously the prerogative of white males.[27] Because CA/T is obviously changing the landscape of the Boston area and concomitantly life lived in this human-built environment, a broad sector of the public, informed by the media, especially the *Boston Globe,* takes an interest in the project. Because of its increased skepticism toward technology and its creators, a large segment of the public and the media currently examines the values and policies of those who preside over the project. Because of the destruction wrought on the environment by previous large-scale projects, the citizens of Boston, with the aid of government regulations, warily monitor the environmental impact of the project.

Public participation and implementation of environmental legislation as exemplified by the CA/T Project suggest a change in the public attitudes toward large-scale technological projects.[28] Political activists especially, from the late 1960s through the 1970s, effectively opposed the initiation of large projects, particularly the construction of highways.[29] The politically and environmentally sensitive system builders who conceptualized the CA/T Project successfully negotiated with this opposition; in so doing they have provided a model for would-be system builders elsewhere. A key characteristic of the new approach is public participation. This implies the need for a public informed about the creative process of system building and the ways in which its values can be embedded in the second creation. Prometheus the creator, once restrained by defense projects sharply focused upon technical and economic problems, is now free to embrace the messy environmental, political, and social complexity of the postindustrial world.

> > > > > > > > > >

II | M I T a s S y s t e m B u i l d e r :

SAGE

As an American institution devoted to education and the advancement of knowledge in the field of technology, MIT has always felt a special responsibility to render public service, especially to any branch of local, state, or federal government.

"Policy Related to Government Contracts" (1944)

▼ MIT assumed this special responsibility wholeheartedly when it became the system builder for the SAGE Project (Semiautomatic Ground Environment), a computer- and radar-based air defense system created in the 1950s. The SAGE Project presents an unusual example of a university working closely with the military on a large-scale technological project during its design and development, with industry active in a secondary role. SAGE also provides an outstanding instance of system builders synthesizing organizational and technical innovation. It is as well an instructive case of engineers, managers, and scientists taking a systems and transdisciplinary approach.

In the opinion of many knowledgeable persons, the SAGE Project was a failure. By the time of its initial deployment in 1958, the intercontinental missile had become the looming threat for United States defense planners, not the long-range bombers against which SAGE had been designed. Yet, SAGE still had notable positive—and unintended—consequences in the realm of computers, communications, and management. The SAGE Project can be compared to the early-nineteenth-century Erie Canal project as one of the major learning experiences in technological history.

The history of the SAGE Project contains a number of features that became commonplace in the development of large-scale tech-

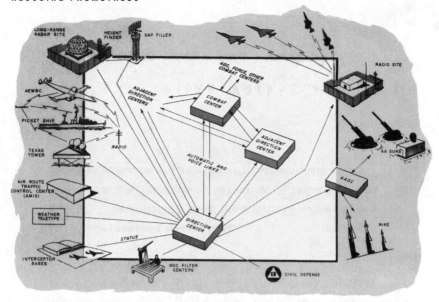

THE SAGE AIR DEFENSE SYSTEM. (Courtesy of MITRE Corporation)

nologies. Transdisciplinary committees, summer study groups, mission-oriented laboratories, government agencies, private corporations, and systems-engineering organizations were involved in the creation of SAGE. More than providing an example of system building from heterogeneous technical and organizational components, the project showed the world how a digital computer could function as a real-time information-processing center for a complex command and control system. SAGE demonstrated that computers could be more than arithmetic calculators, that they could function as automated control centers for industrial as well as military processes.

]]]]]]] The Cold War and Technical Momentum

The Soviet Union's detonation of an atomic device in 1949 caused grave concern in the United States, especially among members of the Atomic Energy Commission, the Department of Defense, and the Truman administration. Decisions to support the development of a fusion bomb and to enhance the striking power of the Air Force followed. The Truman administration assigned the Air Force the responsibility of upgrading a ground-based air defense system that critics described as inadequate, even blind and dumb. The arrival of the era of long-range bombers and missiles drastically altered Americans' long-standing belief that they were insulated from air attack.

The decision to develop a defense system against air attack was prompted not only by these military and foreign-policy developments but by technical developments that set the stage for the development of a radar- and computer-based defense system. In 1948, Karl T. Compton, president of MIT and a science and technology policy adviser to the federal government, asked N. M. "Nat" Sage, director of MIT's Division of Industrial Cooperation, to prepare a report about recent developments at MIT of interest to the military in the fields of guided missiles, radar networks, and gunfire control, and especially the military applications of high-speed digital computers.[1] The items he enumerated as ripe for development became the principal components of the SAGE air defense system.

Sage—the man, not the system—asked Jay W. Forrester, associate director of MIT's Servomechanisms Laboratory, who had suggested to Compton to ask Sage for the report, to reply to Compton. Head of the electronic computer division of the Servomechanisms Laboratory, Forrester responded with a twenty-three-page report about the broad utility of digital computers, especially for military applications.[2]

Forrester proposed a fifteen-year research and development project to cost $2 billion. This figure compares to the cost of the wartime Manhattan Project plus the cost of the various projects of the MIT Radiation Laboratory (Rad Lab), which had developed and deployed radar during World War II. The report asserted that a computer information system for military operations had reached about the same stage of development as radar had in 1937. Forrester hoped that his report would circulate in science and technology policy circles.

]]]]]] The Valley Committee (ADSEC)

Pressured by the Truman administration, Air Force headquarters asked its Scientific Advisory Board (SAB) to consider the need for an improved ground defense against air attack. Headed by Theodor von Kármán, a Hungarian émigré who was a California Institute of Technology (Cal Tech) professor with an international reputation as an aeronautical engineer, the board had been established to utilize the nation's engineering and scientific resources to be found in academia and industry. The Air Force had been separated from the Army for only two years and had not yet developed a substantial in-house research and development program.

Von Kármán established the Air Defense Systems Engineering Committee (ADSEC) (hereafter referred to as the Valley committee) to look into the present state and future development of an air defense system. Conventional accounts of the process of invention, research, and development rarely acknowledge the important role played by committees as contrasted with that of individuals. George E. Valley, Jr., an MIT associate professor of physics who had been at the Radiation Laboratory during the war and had helped edit its seminal technical reports after the war, headed the new committee. As a member of the electronics panel of the Scientific Advisory Board, Valley had taken the initiative earlier to personally inspect the existing radar-based air defense system. Appalled by its primitive character, he had expressed his concerns to the von Kármán committee.[3]

Valley chose a transdisciplinary committee with academic members from physics, electronics, aerodynamics, and guided missiles.[4] Since most members of the Valley committee came from the Northeast, primarily from MIT, Valley suggested that another committee might be needed to provide West Coast defenses.[5] Within several months, the committee had generated a conceptual design for an air defense system and a plan for a research and development program.

The Valley committee began meeting in Cambridge in December 1949. At the peak of its activities, it convened each Friday. This committee began by studying histories of air defense during the Second World War. Surveying existing air defense in the United States, the group discovered a system dependent on human operators who processed data and communicated by voice, face-to-face, or by telephone. Such a system appeared primitive compared even to the machine-dependent air defense systems installed in combat information centers on World War II naval warships.

Sectors of ground air defense covered about several hundred thousand miles each. These had—or were intended to have—five to fifteen large ground-based radars. At the radar stations, operators processed, or "filtered," data from the radar into latitude and longitude coordinates. Each area had an area control center (ACC) with voice telephone communication links to the radar and to ground observers in the field. At these centers, operators calculated aircraft positions, manually plotting them on a large screen, or situation, board. Informed of hostile aircraft in the area, the ACC intercept officer would designate the particular radar station to intercept the bomber; an operator at the radar station would commit an intercepting aircraft, directing the pilot by voice commands to the target.[6] On his evaluation, Valley wrote that "the individual computations were straightforward enough, and anyone could combine the data on a map if he had enough time," but that under attack conditions, humans would face an impossible task.[7]

In a mass attack, the existing air defense system, even if fully implemented, could track or intercept only a small percentage of aircraft and the installed radar could not track low-flying aircraft. After surveying the existing system, committee members found the situation "depressing but not discouraging." The committee proposed replacing the few large radars with several thousand mostly automated (unattended), serially produced small radars able to track both low- and high-altitude aircraft. The committee believed that ordinary telephone lines could be used to transmit coded information from the radars to machine "data analyzers," which would be linked together and communicate to an area command center by telephone lines. The reference to data analyzers anticipated the use of computers, but in its

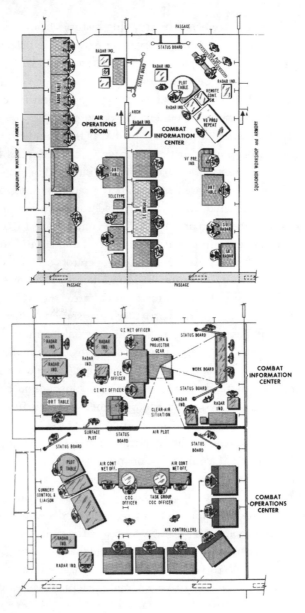

NOTE SIMILARITIES BETWEEN THE COMBAT INFORMATION CENTER OF THE AIRCRAFT CARRIER *PHILIP-PINE SEA* AND THE PROPOSED AIR SURVEILLANCE AND CONTROL ROOM FOR SAGE. *(From Project Charles, Problems of Air Defense: Final Report of Project Charles, Massachusetts Institute of Technology, 1 August 1951, pp. 69 and 73)*

early deliberations the committee had only the haziest notion of what an analyzer would be like.[8]

Members of the Valley committee conceived their task to be the automation of the air defense system. At the time, automation had become a much-talked-about concept, in no small measure because of Norbert Wiener's publication of *Cybernetics; or, Control and Communication in the Animal and the Machine* (1948), a book widely read in the science, engineering, and social science communities. Wiener promoted the possibility of using electronic computation machines (computers) for processing information needed by decision makers in the military, business, and other areas of society.[9]

Like automation, a systems approach to technical problems had a high-tech appeal. Committee members believed the concept of "system" to be generally so unfamiliar that they proceeded to define it for the readers of an October 1950 Valley committee report:

> The word itself is very general . . . [as for instance] the "solar system" and the "nervous system," in which the word pertains to special arrangements of matter; there are also systems of philosophy, systems for winning with horses, and political systems; there are the isolated systems of thermodynamics, the New York Central System, and the various zoological systems.
>
> The Air Defense System has points in common with many of these different kinds of systems. But it is also a member of a particular category of systems: the category of organisms . . . [defined as] "a structure composed of distinct parts so constituted that the functioning of the parts and their relations to one another is [*sic*] governed by their relation to the whole." The stress is not only on pattern and arrangement, but on these also as determined by function, an attribute desired in the Air Defense System.
>
> The Air Defense System then, is an organism. . . . What then are organisms? They are of three kinds: animate organisms which comprise animals and groups of animals, including men; partly animate organisms which involve animals together with

inanimate devices such as in the Air Defense System; and inanimate organisms such as vending machines. All these organisms possess in common: sensory components, communication facilities, data analysing devices, centers of judgement, directors of action, and effectors, or executing agencies. . . .

    It is the function of an organism . . . to achieve some defined purpose.[10]

Ears and eyes become radar; the nervous system resembles a telephone system; human cognition is done by punch-card accounting machinery and possibly digital computing machines; automatic gunfire control replaces human operators; and muscles give way to servomechanisms. At a time before political correctness had reached the engineering and science community, committee members used as examples of a human system one composed of "boss and secretary (low-priced)" and another of "boss and secretary (high-priced)." This contrast associates male bosses and high-priced secretaries with superior powers of analysis and judgment. Pursuing its anthropomorphic metaphors, the committee concluded that the existing air defense system was "lame, purblind, and idiot-like."[11]

To improve the "eyes" of the system, Valley, experienced in microelectronics, organized Thursday evening radar seminars attended by a mix of mathematicians, engineers, and physicists from industry, the universities, and the Air Force. "All these people struggled to invent a new solution to the ground clutter problem," leading Valley to boast that "numerous inventions made in many laboratories and a huge number of journal articles have resulted from this work."[12]

To make the existing system less "idiot-like," the Valley committee decided that the Whirlwind digital computer then being developed at the MIT Servomechanisms Laboratory be used as a data analyzer and that a small pilot system using Whirlwind, several radar stations, and a number of airplanes test the design of a new system, or organism. We shall shortly turn to the uncommon history of Whirlwind.

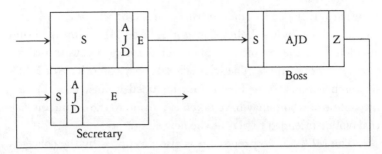

BOSS AND SECRETARY (LOW-PRICED)

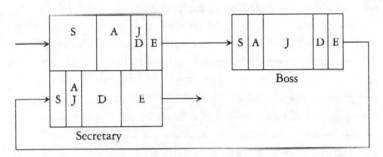

BOSS AND SECRETARY (HIGH-PRICED)

BOSS AND SECRETARY, LOW-PRICED AND HIGH-PRICED: AN EARLY 1950S CONCEPT OF AN INFORMA-
TION SYSTEM. TAKING A NON–POLITICALLY CORRECT STANCE, THE COMMITTEE PORTRAYED THE LOW-
PRICED SECRETARY AS LESS ABLE TO MAKE DECISIONS (D), EXERCISE JUDGMENT (J), AND DO
ANALYSIS (A). *(From Project Charles,* Problems of Air Defense: Final Report of Project Charles, *Massachu-
setts Institute of Technology, 1 August 1951, p. 96)*

]]]]]] Project Charles

At the suggestion of the Valley committee and of Louis Ridenour,
who had recently been named the Air Force's chief scientist, Air
Force Chief of Staff General Hoyt S. Vandenberg asked MIT to
establish a mission-oriented laboratory designed to develop an air
defense system. In a counterproposal, MIT Chancellor Julius Stratton, with the concurrence of MIT president James Killian, proposed that all three branches of the military take part in funding
the new laboratory. The institute did not want the new laboratory

to be entirely dependent on the Air Force and on the air defense system as defined by the Valley committee. The MIT administration also felt that the committee plans for a laboratory were insufficiently comprehensive. The scientific advisory committee of MIT's Research Laboratory of Electronics included influential MIT faculty, some of whom may have taken exception to the dominant role that Valley intended to play at the new laboratory.

The MIT counterproposal was also prompted by the concerns of MIT physics professor Jerrold Zacharias. Seeing the Air Force as inexperienced in science and technology and at the same time as arrogant, he argued that the role of the Air Force in such a laboratory should be tempered by Army and Navy presences. A "young, but cocky Air Force, wedded to the myth of 'Victory Through Airpower,'" Zacharias argued, would concentrate its resources in long-range bombers and atomic weapons rather than on defensive systems. As a result, it would be satisfied with a quick and inexpensive engineering fix that did not take into account the many political and social factors involved in a defense against air attack.[13]

Zacharias contended that air raid shelters, underground military facilities, and dispersal of industry should be considered alternatives or supplements to radar and interceptor defense. He also anticipated the development of intercontinental ballistic missiles and the availability of nuclear bombs, a combination that could overwhelm the envisioned air defense system. The passing through the defense net of only a small fraction of nuclear-armed missiles and bombers would bring intolerable mass destruction unless other defense measures were taken.[14]

To avoid a hasty technical fix, Stratton proposed—probably at the suggestion of Zacharias—a "summer study project," an organizational form that was becoming a hallmark of MIT's research and development style.[15] The project would consider the laboratory proposal broadly and in depth, then prepare a summary report that would decide MIT's role in the establishment of the proposed laboratory.

An organizational innovation suggested by academics, the military-funded summer studies project provided connective tissue between the military and the university. Usually convened during the months when academics were free from teaching responsibilities, summer studies sometimes extended over a longer period. These studies made possible the focus on a single problem by a transdisciplinary group of academics and scientists and engineers from industry. "With participants living and eating under the same roof, the work proceeded very nearly twenty-four hours a day. . . . The intensity of concentration was enormous"[16]

Summer studies have generally been positively evaluated, but in several cases participants came to the reluctant conclusion that the military sponsor intended that the participating scientists legitimate decisions made previously by the military. Some engineers and scientists also felt that the military defined complex problems too narrowly. Zacharias, leader of an early and successful summer study, was known for his insistence that summer study problem definition be broad so that the participants could take a wide-ranging systems approach.[17] Valley took a negative view of summer studies, whether narrowly focused or wide-ranging: "This was pure Zacharias. . . . Zacharias was the inventor of MIT study groups, which consisted of getting a treeful of owls together, furnishing them with a tree, and writing a report—that did or did not have something to do with reality."[18]

Zacharias headed Project Hartwell in 1950, a major summer study that became a model for other summer study projects. The origins of Hartwell can be traced back to early 1950 when Mervin J. Kelly, president of the Bell Telephone Laboratories, advised the chief of naval operations, Admiral Forrest P. Sherman, who was concerned about the vulnerability of U.S. ships to Soviet submarines, "to go to some place like MIT, since they have a lot of screwball scientists who will work on anything, and get a short study made."[19] Zacharias accepted the directorship of Project Hartwell on the condition that the Navy expect the scientists to come up not with a weapon but with some systematic thought about the broader subject of security of overseas transport. Of the thirty-three scientists involved, ten were alumni of the MIT Radiation Laboratory or of the Manhattan Project. This explains why contemporaries sometimes referred to a wartime research-laboratory "Mafia."

Zacharias stressed the systems approach taken at the laboratory. In his view, the Rad Lab was the standard by which all large-scale research and development projects should subsequently be judged. While nine of the Hartwell group were from Harvard or MIT, three came from Bell Laboratories, known then for its competence in systems engineering. Zacharias claims that the Hartwell summer study brought systems engineering to the Navy.[20]

Valley and the Air Force were disappointed at the delay in establishing the research and development laboratory but accommodated themselves to the summer study that was scheduled to begin in February 1951. ("Summer" studies were no longer confined to the summer.) MIT President Killian passed over Zacharias, the obvious choice after his success with Project Hartwell, because the latter did not get along well with Valley and because of Zacharias's outspoken views on the arrogance of the Air Force. The MIT administration and the Air Force picked F. Wheeler Loomis, former associate director of the Rad Lab and chair of the University of Illinois physics department, to lead Project Charles, as the summer study came to be known. (Years later, despite their differences, Zacharias commended Valley for bringing an unenthusiastic Air Force to the point where it would take seriously a broadly conceived ground defense system, in contrast to an air-offensive one.)

Zacharias became associate director of Project Charles. When the summer study group divided into subcommittees, he chose to head the one on "passive defense," which was to emphasize dispersal of industrial and population concentrations, and the construction of air raid shelters. Zacharias brought in economists and social scientists to advise his committee, including Paul Samuelson from MIT, James Tobin from Yale, and Carl Kaysen from Harvard. Zacharias's doubts about the effectiveness of the air defense being defined by Project Charles only increased as the possibility of intercontinental ballistic missiles with nuclear warheads became a reality.

The study group and the consultants represented the leadership—mostly East Coast—of the military-industrial-university community.[21] Of the twenty-eight members, eleven came from

MIT, only five came from other universities. The MIT members included, besides Valley and Zacharias, Gordon Brown, director of the MIT Servomechanisms Laboratory and future MIT dean of engineering; Forrester, director of the Whirlwind computer project and head of MIT Digital Computer Laboratory; Albert G. Hill, physicist and head of the MIT Research Laboratory of Electronics; J. C. R. Licklider, a psychologist and, later, a legendary computer pioneer; and H. Guyford Stever, aeronautical engineer, later a presidential science adviser. Members also came from Hughes Aircraft, Bell Telephone Laboratories, and the Polaroid Corporation, represented by its brilliant head, Edwin Land. Consultants brought in from time to time included Kaysen from Harvard University, Charles C. Lauritsen, a prominent member of the Cal Tech faculty, John von Neumann from the Institute for Advanced Study, and Jerome B. Wiesner from MIT.

Having been briefed by military intelligence officers about the existing threat from Soviet TU-4 bombers and about the possibility that by 1958 supersonic high-altitude long-range bombers and guided missiles would be a threat as well, the Project Charles group quickly recommended integrating the air defense systems of all three services. They then proposed in detail a research and development program for near-term improvements in the existing air defense system and long-range ones for a mostly new system of "aircraft control and warning." Acknowledging the limitations of such a committee of experts in a democratic society, the Project Charles report stated that the extent and expense of the new system should be decided by the political representatives of the people: "It is then the problem of technical people to see that the funds are spent in ways that will be most effective."[22]

The group emphasized the development of a digital computer for information processing and control to be located in an "Information and Control Center" analogous to the combat information centers that had been installed on battleships during World War II. Unlike the Valley committee's vague reference to a data processor, Project Charles recommended the development of a large digital computer that would embody and improve on the best features of the Whirlwind computer then being developed by the Forrester team. The committee recommended a research program to increase the speed, capacity, and reliability of the computer's memory—both the electrostatic tubes then being used for memory and the three-dimensional array of ferromagnets recently invented by Forrester, as well as to develop other computer components, including transistors. So improved, computers would become the "brains" of a centralized, automated air defense system, first called the Lincoln Transition System, then later, in 1953, named the Semiautomatic Ground Environment (SAGE). Once deployed, this system became the first successful effort to "apply computers to large-scale problems of real-time control as distinct from calculation and information processing."[23] The history of Project Charles demonstrates the ability of a committee to play the role of a system builder, at least during the early conceptual and planning stage of a project.

The final Project Charles report called for initiatives in addition to a new air defense system. These initiatives included the development of interceptor aircraft, air-to-air guided missiles, long-range surface-to-air missiles, and a "passive defense" program for dispersal of new industrial construction and air raid shelters.

]]]]]] The Lincoln Laboratory

In September 1951, the Air Force awarded MIT a contract to carry out Project Lincoln, which was designed to conduct research and development for fulfilling many of the recommendations of Project Charles. Project Lincoln also involved the establishment of a related research and development laboratory to be administered by MIT. Louis Ridenour, head of the Air Force Scientific Advisory Board, had importuned Killian to undertake the proposed laboratory and the research and development (R&D) programs, not only because of MIT's responsibility to respond to a national need but also because the laboratory would insure MIT's leadership role in electronics research.[24] Killian probably saw the analogy between the way in which the wartime Rad Lab had propelled MIT into the field of electronics and the possibility that the new laboratory would provide a similar impetus.

THE LINCOLN LABORATORY BUILDINGS IN THE EARLY 1950S. THE SQUARE BUILDING AT LEFT HOUSED THE COMPUTER. *(Courtesy of MITRE Corporation)*

Another reason for MIT's involvement, proffered some years later by Zacharias, was the unwillingness of his friends "up the street" at Harvard to respond to the nation's defense needs. Harvard President James Conant had declared that the university should not become engaged in classified research. MIT, on the other hand, Zacharias contended, did not believe that "the safety of the country should be sloughed off on somebody else because it's dirty work. . . . There is a need for an institution such as this one [MIT] to come to the aid of its country when it's in trouble."[25] In making this argument, Zacharias spoke for many at MIT who did military research and development despite the criticism of some of their colleagues in other universities.

Initially, F. Wheeler Loomis directed Project Lincoln's laboratory, soon thereafter named Lincoln Laboratory. Built at Bedford, Massachusetts, the laboratory was less than an hour's drive from MIT's Cambridge campus and next to the Air Force's Hanscom Field. To Valley's dismay, Loomis agreed to remain only a year; after that year, Albert G. Hill, head of MIT's Research Laboratory of Electronics, replaced him. Valley had felt comfortable with Loomis, but did not with Hill, a friend and colleague of Zacharias. Valley and Hill soon found that they stood miles apart in their approaches to the development of air defense. Hill, for one, felt that Valley's ideas resembled the quick technical fix that Zacharias opposed.[26]

Valley headed Lincoln Laboratory's Division 2, which developed radar and communications. Forrester's MIT Digital Computer Laboratory became Lincoln Division 6, which continued to develop the Whirlwind computer and its successors for use in the air defense system. Breaking down Lincoln into such functional divisions created problems of systems coordination, or systems engineering.

]]]]]] Project Whirlwind

The introduction of electronic high-speed digital computers capable of real-time information processing and of the "command and control" of large-scale complex systems marks the major technological achievement of the SAGE Project.[27] Though SAGE is conventionally portrayed as an air defense system, it can also be described as an information-processing and real-time control system. Conceived in this way, the development of the Whirlwind computer and its production derivatives, the AN/FSQ-7 and AN/FSQ-8 computers (Army-Navy Fixed Special Equipment), becomes central to our understanding of SAGE's achievements.

The background for the decision to use the Whirlwind computer as the air defense information processor extends back to January 1951, about a month before the establishment of Project Charles. In a chance conversation with Valley about the deliberations and problems of the Valley committee, Wiesner, then an MIT professor, suggested to Valley, who sought a data processor for the projected air defense system, that he visit the Whirlwind computer being developed by Forrester and his associates.

The Navy had funded the MIT Servomechanisms Laboratory in 1944 to design and develop an analog computer to function as an airplane stability and control analyzer. Its function was to simulate the effect of various airplane controls on airplane performance and to perform "rather like an elaborate aircraft pilot trainer . . . precise enough to take wind tunnel data from a model of a proposed plane and predict the behavior of the full-scale airplane before construction."[28]

Forrester, with his close engineering associate Robert R. Everett, soon envisioned goals for Whirlwind far beyond airplane simulation. Forrester was guided by the crucial belief that creative inventors and developers should not be constrained by early concepts when ongoing development of a device reveals heretofore unseen possibilities of surpassing importance.[29]

Early in the Whirlwind project, Perry Crawford, Jr., a friend of Forrester's and an MIT graduate who acted as liaison between the Naval Special Devices Center and those working on projects that the center sponsored, suggested that Forrester develop a digital rather than an analog computer. Encouraged by the early problem-solving successes of the ENIAC (Electronic Numerical Integrator and Computer), a large-scale, general-purpose digital machine built at the University of Pennsylvania during the war, Forrester decided to shift from the analog to the digital approach with which engineers were less familiar.

The designers of the project, now named Whirlwind, quickly began to envision a general-purpose, information-processing computer that could automate many of the functions previously performed by humans in World War II shipboard combat information centers to coordinate defense against air attacks. The combat information centers took surveillance information from radars and directed intercepting aircraft, but human operators, not computers, did most of the information processing. To conceive of digital computers as general, rather than special-purpose, machines enabled many pioneers in the computer field to discover new computer applications.

Aware of Whirlwind's background history, Wiesner informed Valley that the Office of Naval Research, which had taken over funding of Whirlwind, was unhappy about rising Whirlwind costs and delays. Forrester wanted to spend at the rate of $100,000 a month, which would grossly exceed the annual ONR appropriation of $750,000.[30] Forrester's ingrained tendency to think like a pragmatic, problem-solving engineer did not fit in well with the mathematical bent of Dr. Mina S. Rees, who headed the ONR Mathematics Branch. She, like many other mathematicians and scientists, conceived of a computer as a scientific instrument suited for research, not as a general-purpose device for information processing and process control.[31]

After his conversation with Wiesner, Valley questioned MIT colleagues about the Whirlwind project. He got mostly negative opinions to the effect that it was out of step with sound, mainstream computer development. Valley's initial impression of the partially built computer already using eight thousand vacuum tubes, a monster machine he could walk around in, was guarded but favorable. He considered Whirlwind mechanically overdesigned but felt relieved to find Forrester and Everett open to suggestions for design changes.[32] On the other hand, he took them to be annoyingly cocky when they presented their sales pitch. "They'd been put in charge of this and nobody'd ever reined them in. . . . They were just off on their own spending money like it was out of style."[33]

Valley opted in favor of the Whirlwind for the air defense project, possibly because he had no other choice. Robert Buderi, in his book about SAGE, reports a conversation between John Marchetti of the Air Force Cambridge Research Laboratories and Valley:

"George, do you really want that kludge?"
"Well, have you got something else?"
"No."[34]

"When the word spread," Valley adds, ". . . that I had gotten into bed with Whirlwind, a number of busybodies warned me it was a grave error."[35]

After becoming Division 6 of the Lincoln Laboratory, the Whirlwind Project continued to concentrate on solving the critical problem of developing a high-capacity, real-time, random-access memory for the Whirlwind and successor computers. Without this, computers could not track a sizable number of aircraft and interceptors.[36] Forrester's description of the essence of the computer merits summary: the digital computer is a machine consisting of four principal components: the terminal equipment, including punch-card machines, magnetic and paper tape equipment, typewriters, and printers by which operators communicate with the computer; the arithmetic element, which he compared to a desk calculating machine; the internal control unit, whose function he compared to the role of the operator of the desk calculator; and the memory, which stores information, including central instructions and data, or information.

Of the principal components, Forrester identified memory as the critical problem, or reverse salient, in the developing computer front. He found the other components more advanced because their postwar development had drawn heavily on the progress made in the radar field during World War II, especially at the Radiation Laboratory at MIT. Radar provided the basis for the electronic circuits adapted for use in the arithmetic units and the control units of the digital computer. Radar also refined the pulsed-circuit video techniques used for coding information in the computer. (The pulses served a function comparable to that of Morse code in the telegraph.) In summary, Forrester testified that all components of the digital computer except the memory "could be developed in a fairly straightforward way out of the electronic circuit technology that had been left at the end of World War II."[37]

Experiencing disappointing results after having tried magnetic drums and acoustic delay lines, Forrester felt the most likely candidate for the memory component was electrostatic storage tubes, but he found himself nursing these "balky storage tubes late into the night."[38] His dissatisfaction with storage tube reliability and cost—and the resulting search for an alternative—led him to invent the magnetic flux core memory. By correcting a reverse salient in an expanding system, Forrester made a milestone advance in the history of computing.

Before turning to the core memory, we should consider Forrester's inventive style. This can be characterized as a lifelong effort to develop and understand feedback-rich dynamic systems. The similarities between Forrester's style and that of Elmer Sperry (1860–1930) are uncanny. Sperry is the outstanding American pioneer in the introduction of feedback controls for ships, aircraft, and other technical systems early in this century.

Both men had similar backgrounds. Sperry, like Forrester, grew up on a farm; Sperry near Cortland, New York, and Forrester in the Nebraska sand hills.[39] Both their mothers taught school. As children, both Sperry and Forrester experimented with "batteries, doorbells, and telegraphs." Guidance and inspiration came for Sperry from reading popular mechanics books in the local YMCA library; for Forrester, from the Nebraska "traveling library" that lent books to local schools. As they attended local high schools, both built electric supply systems—Sperry for the local church, Forrester to bring electricity to his parent's ranch. Sperry, however, attended a normal, or teacher's, college at a time when engineering education was not commonplace; Forrester earned both a bachelor of engineering degree (University of Nebraska, 1939) and a master of electrical engineering degree (MIT, 1945) at a time when formal engineering education had become commonplace. His experience as an MIT graduate student in electrical engineering at a time when the subject was on the front edge of engineering advance forever conditioned him to conceptualize in terms of circuits, systems, and feedback loops.

Both Sperry and Forrester were feedback solutions searching for problems. Sperry spoke of using feedback controls for "taming the beasts." He moved from less to more complex electromechanical systems; Forrester ventured from technical systems to extremely complex social systems. With funding from the military, both were able to delve deeply into feedback controls. Sperry became an outstanding engineer in the context of a World War I military-industrial complex; Forrester flourished in a World War II–spawned military-industrial-university complex. Each chose to solve the complex problems that, as Sperry said, allowed them to avoid the vulgar competition; each had "the courage and persistence to sustain a long-term vision against oppositions that arise along the way." And each chose to leave a particular field of inventive activity after having made fundamental contributions. "Each change," Forrester observes, "led to pioneering in new and more challenging areas."[40]

Forrester has summarized his inventive style:

> There are many inventions that are original, clever, impractical, and useless. I have always started from a clear need and objective and searched for ways to meet the challenge. Such was true in the early days of computers for military combat information centers, in seeking a better information storage for computers that led to the random-access memory, in developing the field of system dynamics as a powerful way to design corporate policies, in departing from the conventional analysis methods of the social sciences in my "Urban Dynamics" and "World Dynamics" books, and most recently with system dynamics as a fundamentally stronger and more realistic foundation for K–12 education.

We shall consider Forrester's inventions in the field of system dynamics later, but now, in the case of his search for the random-access memory, we find him contemplating magnetic materials that he believed offered one possibility, albeit one seemingly remote. These materials could be magnetized and demagnetized by pulses of electricity so that their state of magnetization or nonmagnetization could indicate OFF or ON. OFF and ON patterns could represent information. A major challenge in developing a magnetic memory was to find a material that could reliably maintain an ON or OFF state and, more important, could switch states rapidly.

A 1949 article in an electrical engineering journal that described the magnetic properties of Deltamax caught Forrester's attention. The Germans had used this magnetic material in World

War II to make analog control devices for tank turrets.[41] A graduate student working in the laboratory recalls that in the fall and winter of 1949–50

> Jay took a bunch of stuff and went off in a corner of the lab by himself. Nobody knew what he was doing, and he showed no inclination to tell us. All we knew was that for about five or six months or so, he was spending a lot of time off by himself working on something. The first inkling I got was when he "came out of retirement" and put Bill Papian to work on his little metallic and ceramic doughnuts.[42]

Papian, and numerous other graduate students in MIT engineering then, worked on projects in MIT research laboratories as part of their master's or doctor's thesis. Both Forrester and Everett had done their masters' theses in the Servomechanisms Laboratory. Kenneth Olsen, a one-time graduate student who worked in the Forrester laboratory and later founded the Digital Equipment Corporation, a computer manufacturer, recalls: "We were cocky. Oh, we were cocky! We were going to show everybody! And we did. But we had to lose some of the cockiness in the sweat it took to pull it off."[43] Some MIT professors also found Forrester and Everett cocky in their self-assurance, for neither had a doctorate and neither was a professor or a lecturer.

Forrester first asked Papian to test Deltamax cores made by Allegheny Ludlum. Later his attention shifted to small doughnut-shaped magnetic ferrite cores made by a German ceramicist whom Forrester had located in New Jersey. Supervised by him, the ceramicist slightly varied the material composition of the cores, seeking a combination with the magnetic characteristics, including a square hysteresis curve, that would provide lightning-fast magnetization and demagnetization. Because the magnetic properties of the material did not vary in a linear fashion, the Forrester team used a cut-and-try approach.

As early as October 1949, Forrester and Papian's cores tested out many times faster in their magnetic reversals than those ini-

JAY FORRESTER HOLDING A CORE MEMORY COMPONENT, CIRCA 1970. *(Courtesy of MITRE Corporation)*

tially tested, but not until 1952 did Forrester confidently predict that a matrix magnetic-core memory would fulfill the needs of Whirlwind.[44] The experimentation, testing, development, and manufacture of the memory cost more than $1 million. If the development had taken place under commercial circumstances rather than with the lavish funding of the Air Force, completion might have been delayed for months, even years.[45] The memory became a part of the AN/FSQ-7, a production version of the Whirlwind used in the SAGE system. In 1955, International Business Machines (IBM) introduced the core memory on its commercial 704 machine.

Other engineers and scientists outside MIT made contributions to the development of memory, but Forrester made the critical contributions that brought assured widespread adoption. He saw the need for a three-dimensional, rather than a two-dimensional, array, or matrix, in order to greatly increase the capacity of the memory unit. He also found a way of accessing and reading the memory without destroying the magnetic patterns. Like so many other inventors, Forrester proceeded by analogy: having found that a three-dimensional matrix of electrostatic storage tubes increased memory, he tried this architecture with the core memory.

As was the case with many other major inventions, a number of persons were attempting, at about the same time, to solve the same critical problem, or correct a reverse salient, in an advancing technical front. An Wang, a scientist at Harvard University, for instance, used Deltamax not in a three-dimensional but in a one-dimensional array, as a method of storing digital information. Jan Rajchman of Radio Corporation of America, J. Presper Eckert and Chuan Chu of the Moore School of Engineering of the University of Pennsylvania, F. W. Viehe, and Munro Haynes, a graduate student at the University of Illinois who carried his thesis work on magnetic memory to IBM as an employee—all figure in the history of the development of computer memory.

Because computer manufacturers, including IBM, adopted and commercialized magnetic-core memory in the late 1950s, the virtually inevitable patent litigation ensued. The Radio Corporation of America (RCA), IBM, and MIT became entangled in a long and expensive patent dispute in the late 1950s, a case eventually settled out of court when IBM agreed to pay MIT a one-time licensing fee of $13 million. This sum, only a fraction of what MIT sought, turned out to be an even smaller fraction of the value of core memories to IBM and the computer industry in the next two decades.[46]

Looking back on the reception—and the nonacceptance—of the magnetic-core memory, Forrester recalls: "It took us about

AN ARRAY OF FERRITE CORES. *(Courtesy of MITRE Corporation)*

seven years to convince industry that [random access] magnetic-core memory was the solution to a missing link in computer technology. Then we spent the following seven years in the patent courts convincing them that they had not all thought of it first." A complaint not uncommon among inventors.[47]

Other features besides the memory help explain the recognition now of Whirlwind as a milestone in the history of computing. Taking information directly from the radar sensors and processing it immediately instead of taking stored information in batches from punched cards or other input devices, Whirlwind has claim to be the first major real-time control computer.[48] Forrester and his associates also found means to greatly reduce vacuum tube failure by increasing their life by several orders of magnitude and finding a method for identifying deteriorating tubes before they failed. They introduced an early version of time-sharing by which several monitors share time on a single computer. Cathode-ray tube displays of computer information also became a feature of Whirlwind and its derivative computers, including the production versions used in the SAGE system, the AN/FSQ-7 and AN/FSQ-8. In addition, development of an unprecedentedly large software program to provide the operating system for the AN/FSQ-7 also became a milestone in the history of programming.[49]

]]]]]] The Michigan Alternative

In December 1951, almost six months after MIT had inaugurated Project Lincoln and initiated plans to build Lincoln Laboratory to house the air defense projects carried out under the umbrella of Project Lincoln, President Killian wrote to Secretary of the Air Force Thomas Finletter informing him that Project Lincoln projected a large budget over the next few years. Loomis, director of Project Lincoln, had warned Killian that the expensive and long-term endeavor could fail and embarrass MIT unless generous and long-term support were guaranteed. Killian told Finletter that Project Lincoln not only drained human resources from teaching and fundamental research but would place MIT in financial jeopardy if the government delayed funding. Killian intimated that MIT might withdraw from the project if assurances were not forthcoming.

In February 1952, Finletter replied that the Air Force "heartily" endorsed and would support the air defense activities of Project Lincoln. He stressed that the Air Force, dependent on eminent scientists at MIT, recognized the heavy burden that such a large-scale engineering and applied-science project placed on the institution. He assured Killian that the Air Force, with some assistance from the Army and the Navy, would provide the approximately $20 million needed in fiscal 1952 for various Lincoln projects. These included the Lincoln Transition Air Defense System later known as SAGE. Loomis predicted that Lincoln's budget would subsequently level off at approximately $20 million annually, of which about $10 million would be earmarked for the Lincoln Transition System.[50]

In January 1953, several months before the first major field test of the Lincoln Transition System that involved the Whirlwind computer and fifteen radar sites—and just a few weeks before Dwight D. Eisenhower's inauguration as president—Killian wrote again to Finletter. As before, Killian raised the question as to whether or not MIT should continue with Project Lincoln. But this time, rather than emphasizing funding, he requested a technical review of Project Lincoln. "If the conclusion is reached that some agency other than MIT should continue to be the contractor for the project," Killian stated, "we stand ready to withdraw."[51] Once again, he stressed that MIT wished to be involved in such a financially hazardous and resource-draining endeavor only if participation was clearly in the national interests.

Killian intimated the possibility of withdrawal in order to pressure the Air Force to decide between the Lincoln Transition System and a competing air-defense project funded by the Air Force at the University of Michigan. A participant in, and historian of, the SAGE Project believes that Killian finessed the situation by saying in effect that "he didn't want to waste his time playing second fiddle to Michigan. . . . This put the prestige of MIT in competition with the University of Michigan."[52]

In his response, Finletter acknowledged that other technical groups were working "diligently" on improvements to the present air defense system and that he felt duty-bound to support them fully. From them, he predicted, might come a system that could substantially strengthen air defense before "the revolutionary LINCOLN system materializes in its entirety."[53]

Two weeks later, Lieutenant General E. E. Partridge, head of the Air Force Research and Development Command, which had overarching responsibility for Project Lincoln, made it categorically clear that the Air Force would not commit itself to the Lincoln Transition System to the exclusion of other defense systems being developed. Using language more common in the military hierarchy than in academic circles, the general wound up a lengthy statement of the facts of the situation—as he saw them—with the enjoinder that "to carry out the . . . [air defense] program in the shortest possible time, the maximum of cooperation and coordination between all agencies [MIT and other] concerned will be required."[54] Until further testing of prototype devices and field trials of small-scale systems had shown the clear superiority of one system over the other, both would receive funds for research, development, and early production. He sent a similar letter to the University of Michigan.

Since January 1951, the University of Michigan's Willow Run Research Center had been working on an Air Force contract to expand a guided-missile program into the air defense system later known as the Air Defense Integrated System for Surveillance and Weapon Control (ADIS). Even earlier, Michigan and the Boeing Aircraft Company had been working together to develop the BOMARC, a defensive ground-to-air antimissile missile named for Boeing and the Michigan Aeronautical Research Center. By 1953, ADIS had evolved into a system that integrated the BOMARC missile, manned interceptor aircraft, antiaircraft artillery, a radar ground-control system, ground observers, and data-processing facilities.[55] General Partridge tersely characterized the ADIS system as decentralizing data processing and weapon control and centralizing threat evaluation facilities, and the Lincoln Transition System as centralizing data processing, weapon control, threat evaluation, and weapon assignment.[56]

At one level, the MIT and Michigan laboratories competed; on another level, Air Force commands competed. The Air Force Air Development Center at Rome, New York (RADC), had responsibility for the "engineering functions" associated with ADIS, while the Air Force Cambridge Research Laboratories at Cambridge, Massachusetts, had similar responsibility for the Lincoln Transition System. Engineering functions included procurement from manufacturers of systems components designed by the university laboratories in consultation with the manufacturers. To complicate matters further, the Air Defense Command, based at the Ent Air Force Base in Colorado Springs, favored the ADIS project.[57] Because of the economic stimulation that large military contracts brought to a region, congressmen from Massachusetts and Michigan entered the competition as well.[58]

Three months later, General Partridge dramatically altered Air Force policy:

> . . . for reasons which will not be enumerated here, the Air Force has found it necessary to revise the aforementioned policy, as outlined in my letter of 28 January 1953, and to initiate a unilateral approach to the solution of its R&D problems in the field of Air Defense Electronic Environments. This single approach will be oriented toward the Lincoln Laboratory Transition Air Defense System, and the ADIS program at RADC will be phased out completely as expeditiously as possible.[59]

Partridge's letter did not offer reasons for his decision. The authors of the authorized history of Project Whirlwind, Kent Redmond and Thomas Smith, affirm that "there were the unavoidable mutterings that the decision had been politically motivated, and it would be naive to assume that political and economic pressures played no role."[60] Valley recalls that eminent scientists, engineers, and industrialists with influence in military, industrial, and university circles supported the Lincoln approach, among them Mervin Kelly, head of Bell Laboratories, Thomas J. Watson, Jr., head of IBM, and Robert C. Sprague, president of Sprague Electric. "The generals at the very top of the Air Force, who respected those men," Valley asserts, "also respected and supported us. But you can't win a battle with generals alone. . . ."[61]

Nonetheless, Partridge's letter did not close the issue. In the spring of 1953, a committee led by Kelly heard testimony that "the Lincoln system was far ahead of its competitors as far as hardware and field testing were concerned," then presented a report to the secretary of defense supporting the decision that favored the Lincoln Transition System.[62] Following the Kelly report, Air Force officers toured both the Lincoln and the Michigan laboratories. At Lincoln, they saw cathode-ray tube displays of actual fighter planes directed by the Whirlwind computer intercept bombers, and they had a tour of the Whirlwind computer, which occupied two floors of the Barta Building at MIT. On the tour at Michigan, the officers saw a demonstration run by a large analog simulator. A pen representing the guided missile intercepted the pen representing the attacking bomber, all of which appeared on a huge plotting board. A number of the officers favored the use of guided missiles, preferring the representation on the plotting board to Lincoln's cathode-ray tube. Although a contingent of young officers at Lincoln understood and highly approved of digital computing, some of the officers representing the Air Material Command and Air Defense Command felt far more at ease and familiar with the Michigan analog computers than with the more recently developed digital one.

Valley found the officers who favored the Michigan system technically naïve. "I was able to keep my temper," he recalls, "and not ask how the hell we were supposed to show the tracks of hundreds of aircraft on a pen and ink plotter without so messing up the paper that you couldn't see anything." He also told the officers that the Lincoln Transition System would take on the problem of directing an interceptor missile when there was one available to test. "I

THE BARTA BUILDING AT MIT. THE WHIRLWIND COMPUTER OCCUPIED 2,500 SQUARE FEET ON THE SECOND FLOOR. *(Courtesy of MITRE Corporation)*

was astonished," Valley adds, that a "trivial exercise in prepro-grammed curve plotting had impressed the majority of the officers as much as had the real thing shown them at Lincoln."[63] Valley seems to have forgotten that many influential Air Force officers were determined to keep the air defense budget small so as to be able to concentrate funds on air offense.

Forrester recalls that he and his Lincoln associates did not believe that the Michigan system could fulfill its promise—in part, because its sponsors asked for too little funding by a factor of ten; and partly because they promised too early a delivery. (Forrester had already become known for thinking big technically and financially.) By early 1953, Forrester continues, it was clear to anyone who com-pared the progress made by the competing systems that Lincoln, though expected to require a longer development period, had moved further along.[64] Others also insist that early field tests of the Transition System in the spring of 1953 demonstrated the validity of the concept, while the Michigan design had "not been demon-strated as successful."[65]

INSTALLATION OF THE COMPUTER IN THE BARTA BUILDING AT MIT. *(Courtesy of MITRE Corporation)*

Transition System field tests took place during the spring and fall of 1953. Engineers working on the Lincoln project had deployed a test version called the Cape Cod System. They used the Whirlwind computer still located on the MIT campus with experimental long-range radar located at the tip of Cape Cod in South Truro, Massachusetts, plus several short-range, gap-filler radars situated around New England. This system successfully located and tracked airplanes designated as enemy bombers, then directed interceptor aircraft to these targets.[66] Michigan could not provide such an experimental demonstration, only simulations.

Valley offers a lively anecdote that hammers home the ultimate Lincoln Transition success.[67] In the fall of 1953, the British Royal Air Force invited U. S. Air Force generals, Valley, Lincoln Project head Hill, and the head of the Michigan project to the south of England for an air defense conference. Valley and the

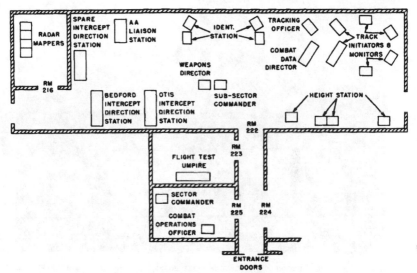

FLOOR PLAN OF THE CAPE COD DIRECTION CENTER. *(Courtesy of MITRE Corporation)*

Michigan representative described their respective air defense concepts. Valley recalls that the Michigan head

> was obviously awed by all the brass, especially by the British superbrass, those sirs and milords who were also professors, air vice-marshals, and so on. He had not learned, as I had learned . . . that in such a situation you stuck a cigar in your face, blew smoke at the intimidating crowd, and overawed the bastards.[68]

Deciding that the Michigan representative "had only the vaguest ideas about physics, electronics, aerodynamics—apparently anything technical," Valley became concerned that the British superbrass, who were snickering and whispering as the Michigan man spoke, would take Americans to be scientifically illiterate. Losing his temper, he rose and thundered, "Shut up, sit down," which, according to Valley, the Michigan spokesman did. After Valley's presentation, he heard the U.S. commander of the Air Defense Command say to the commander of the Air Research and Development Command, "I'm sure glad you picked the right system."

WHIRLWIND PROJECT ENGINEERS, 1947. JAY FORRESTER AT TABLE, RIGHT, WITH BOOKLET, AND ROBERT EVERETT, ASSOCIATE DIRECTOR OF THE WHIRLWIND PROJECT, UPPER RIGHT CORNER OF SAME TABLE. *(Courtesy of MITRE Corporation)*

]]]]]] International Business Machines Corporation

In the spring of 1952, Forrester decided that the time had come to line up an industrial company as subcontractor to participate in design studies for the prototypes of the production model of the Whirlwind and later to manufacture the production model. Traditionally, MIT, though willing to engage in research and limited development, avoided competing with the corporate world in the realm of manufacture. Forrester and Everett visited International Business Machines Corporation, Raytheon Manufacturing Company, and Remington Rand, the three companies considered as possible manufacturers of the production computer. Forrester and his associates had selected these three, along with Bell Telephone Laboratories and Radio Corporation of America, from fifteen to twenty likely companies, a surprisingly large number considering the early stage of the digital computer-manufacturing industry. Bell and RCA, however, withdrew from consideration after deciding that existing commitments did not allow them to take on a project of such magnitude.[69] Because no single company dominated the digital computer industry, as IBM would within a few years, the choice was difficult.

The effectiveness of company organization played as large a role in their choice as did technical competence. Forrester and his colleagues visited the facilities of the three contenders, interviewing engineering, scientific, and managerial personnel. They took especial note of an organization's "degree of purposefulness, integration, and esprit de corps." Forrester probably believed that such characteristics went a long way toward explaining the successes of his own Division 6. They searched especially for evidence of close ties among research, development, production, and field maintenance functions and personnel. The Lincoln engineers also wanted to know about the computer production experience of the companies.

They found that IBM in its Poughkeepsie, New York, factory produced each month digital computing equipment incorporating about 60,000 vacuum tubes. Because the factory "was a closely integrated development and production organization, where the transition from research and development to production had been well established," Forrester was favorably impressed. By contrast, the absence of substantial production capacity at Remington Rand's Eckert-Mauchly division and at its Engineering Research Associates division in St. Paul, Minnesota—the likely sites for the production of the digital computer—counted against that company.

With regard to technical competence, Forrester and Everett found the IBM engineers technically greatly superior to other contestants in the design and manufacture of reliable vacuum tubes and other electronic components of interest to Lincoln. At the time, IBM had 3 million vacuum tubes operating in its equipment in the field. Remington Rand's division, having concentrated on the limited production of large-scale digital computers, did not offset IBM's greater experience with series production, in the opinion of the Lincoln engineers. (By this time, Eckert-Mauchly had the fourth and fifth innovative large-scale UNIVAC computers under construction.)

In his letter summarizing findings and recommendations, Forrester ranked IBM, Remington Rand, and Raytheon in that order and added incisively that "there existed no doubt of proper order of preference." Raytheon's lack of digital computer experience, except in components such as cathode-ray indicators, precluded serious consideration of the company. Remington Rand disputed the selection of IBM as the design and production contractor. One Remington Rand executive asserted that his company had been given a "dirty deal" and had been "unfairly discriminated against by MIT."[70]

Selected as contractor, IBM in 1953 assigned three hundred engineers, scientists, and administrative people to Project High.[71] They participated in the design and manufacture of the AN/FSQ-7, computers that would be installed in the data-processing centers of the air defense system. Initial plans called for the manufacture of about fifty computers, an unusual commitment at a time when producers of such large digital computers usually built one of a kind in a model shop instead of engaging in serial production.[72]

A vacuum tube computer based on the Whirlwind configuration, the AN/FSQ-7 was to have a ferrite core memory even though IBM engineers considered the use of the newly developed core memory problematic. "We all shared Forrester's optimism and arrogance to some extent," an IBM engineer recalls, "yet I wondered at the self-confidence that man must have to convince military leaders to entrust the nation's air defense to such a grandiose collection of basically untried and untested technologies."[73]

Forrester intended his Lincoln team to have ultimate control over the cooperative design and development of the production computer, so he insisted on the right of technical supervision of the cooperative venture and the authority to do the signing off on the final drawings before IBM began manufacture. He also wanted IBM to take responsibility for the performance of the computers after installation. From his early days as a research assistant at MIT's Servomechanisms Laboratory during World War II, he had learned that the engineer should carry his or her responsibilities from the drawing board into the field. He had supervised shipboard installation of devices that he had designed, taking responsibility for subsequent repairs and improvements of them. On the other hand, manufacturers of computers, such as IBM, customarily had not followed up on problems once the product had left the shipping platform.[74]

Forrester insisted upon regular exchanges of ideas and people between his Division 6 and the IBM team. During 1953 and 1954, his engineers sent 1,000 technical documents to IBM and IBM sent 400 to Lincoln. During the first nine months of the project, IBM personnel spent 950 days at Lincoln learning about the Whirlwind and its function in the air defense system.[75] To grind out technical decisions, IBM and Division 6 engineers regularly met in Hartford, Connecticut, midway between Lincoln Laboratory and the IBM headquarters at Poughkeepsie. Their meetings became known as "Project Grind."

Installed in the air defense control centers, each AN/FSQ-7 computer had 25,000 vacuum tubes and two main memories containing 147,456 ferrite cores. Installed in pairs, the giant computers occupied 40,000 square feet of floor space and used 3 million watts of power. Division 6 and IBM took pride in the reliable performance of the installed computers; during their years of service, they lost less than two hours annually of "air defense time."[76] In retrospect, Forrester praised IBM for its participation in the design and manufacture of the AN/FSQ-7s:

> I think the remarkable thing is the broad support that the company gave to the program: the willingness of the company to take what would be seen as risks to make it successful, building a factory, for example, before there was a contract for the product . . . to go ahead even in the absence of certainty of financial remuneration was indicative of the kind of top level support that helped a great deal to bring the whole thing off successfully.[77]

Emerson Pugh, an IBM engineer who has written about IBM's history, adds that the project, while a major gamble that absorbed much of the company's technical resources in an ill-defined electronic data-processing field, paid off in providing a half-billion dollars of revenue and, more important, served as an invaluable learning experience for a generation of IBM people who worked on the first large, real-time on-line computer.

]]]]]] Architecture of the System: The Red Book

The goal that Forrester and Valley now addressed was how to assure the harmonious functioning of the digital computer and search radars. Together they circulated in 1953 a defining technical memorandum, "A Proposal for Air Defense System Evolution: The Transition Phase."[78] In 1955, they supplemented that document with another, which came to be known as the "Red Book." These became the SAGE "bibles."[79] They describe the essential SAGE configuration, or architecture, that survived despite numerous modifications during development of the system.[80]

The SAGE air defense system, as the bibles made clear, would become part of a larger complex presided over by the Joint Chiefs of Staff's Continental Air Defense Command. Other air defense systems in the complex included the Distant Early Warning System, under construction by Western Electric Company; and the Continental Air Defense System being installed by the Bell Telephone Laboratories as an interim, short-term improvement in the system inherited from World War II. Further, the Air Defense Command limited the responsibility of Lincoln Laboratory to the design and development of the semiautomatic processing of the data and weapon-control aspects of SAGE. Responsibility for the deployment of radar, communications networks, interceptor weapons, and other SAGE subsystems fell to other Air Force contractors. The dividing of these functional responsibilities introduced problems of overall system coordination during development and deployment.

SAGE was intended initially to protect the vulnerable northeastern area of the country, although the Air Force planned eventually to deploy it over the entire United States. The planners divided the United States into eight SAGE sectors, each with a combat center equipped with an AN/FSQ-8 computer featuring a specially

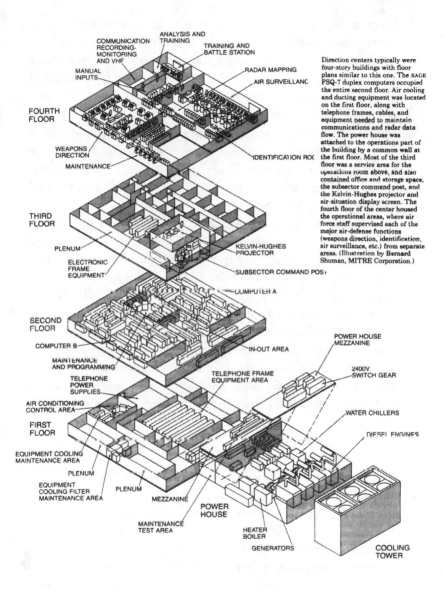

Direction centers typically were four-story buildings with floor plans similar to this one. The SAGE FSQ-7 duplex computers occupied the entire second floor. Air cooling and ducting equipment was located on the first floor, along with telephone frames, cables, and equipment needed to maintain communications and radar data flow. The power house was attached to the operations part of the building by a common wall at the first floor. Most of the third floor was a service area for the operations room above, and also contained office and storage space, the subsector command post, and the Kelvin-Hughes projector and air-situation display screen. The fourth floor of the center housed the operational areas, where air force staff supervised each of the major air-defense functions (weapons direction, identification, air surveillance, etc.) from separate areas. (Illustration by Bernard Shuman, MITRE Corporation.)

Labels on illustration:

FOURTH FLOOR

COMMUNICATION RECORDING-MONITORING AND VHF
ANALYSIS AND TRAINING
MANUAL INPUTS
TRAINING AND BATTLE STATION
RADAR MAPPING
AIR SURVEILLANCE
WEAPONS DIRECTION
MAINTENANCE
IDENTIFICATION ROOM

THIRD FLOOR

PLENUM
ELECTRONIC FRAME EQUIPMENT
KELVIN-HUGHES PROJECTOR
SUBSECTOR COMMAND POST

SECOND FLOOR

COMPUTER A
COMPUTER B
IN-OUT AREA
MAINTENANCE AND PROGRAMMING
TELEPHONE POWER SUPPLIES
TELEPHONE FRAME EQUIPMENT AREA
POWER HOUSE MEZZANINE
2400V SWITCH GEAR

FIRST FLOOR

AIR CONDITIONING CONTROL AREA
EQUIPMENT COOLING MAINTENANCE AREA
PLENUM
EQUIPMENT COOLING FILTER MAINTENANCE AREA
PLENUM
MEZZANINE
POWER HOUSE
MAINTENANCE TEST AREA
HEATER BOILER
GENERATORS
WATER CHILLERS
DIESEL ENGINES
COOLING TOWER

PLAN FOR THE SAGE DIRECTION CENTER. *(Courtesy of Mitre Corporation)*

designed display system, a variation upon the AN/FSQ-7. Combat centers would receive, process, and evaluate information from radar installations in thirty-two subsectors, each with a direction center housing an on-line AN/FSQ-7 computer as well as an identical standby computer. Over telephone lines, long-range radar installations, along with unmanned gap-filler radars that covered areas missed by long-range sweeps, were to send information concerning the location and velocity of aircraft to the computers. Some of the radar would be placed in huge steel platforms called Texas towers, located in coastal waters. Information from Texas towers traveled by microwave radio.

The combat centers were to generate an overall view of threats and responses while the direction centers would process surveillance information about enemy and friendly aircraft, producing information to assign and direct defensive aircraft and missiles. Direction center computers were designed to track a total of two hundred enemy aircraft and two hundred defensive aircraft and missiles. The construction and deployment schedule for SAGE combat and direction centers depended on the rate at which IBM produced the computers. Assuming an output of one computer each month after about two years of lead time, IBM projected November 1960 as the completion date for installation of the full, nationwide SAGE system.[81]

Following the precepts of Project Charles, SAGE designers used the computers to introduce a high degree of automation into the system. No longer would humans scan and analyze radar information, passing it on to others by voice communication, nor would they be the principal information processors, as had been the case in World War II. The primary responsibility for humans in the SAGE system would be their interaction with computers through the use of keyboards and other devices in order to specify which of the airplanes picked up and followed by radar and shown on the computer cathode-ray monitors should be targeted. The computer would then calculate the path that an intercepting fighter or missile should follow in order to destroy the target. Formerly, humans had calculated the path of intersection.

]]]]]] Computer Software Programs

Development of SAGE gave an enormous boost to computer programming which, at the time, had received little attention compared to that given to the designing of the computer hardware. Engineers tended to believe that programming, by means of software, the general-purpose computers for the tasks that they would have to perform would follow easily upon the completion of the hardware.

The classic instance of engineers demeaning programming involves the women who programmed the ENIAC when this early digital computer went on-line in 1946 to prepare ballistic trajectory tables for artillery fire and bombing runs. Dozens of women, many with mathematical backgrounds, had patiently calculated the ballistic tables, as they tediously solved differential equations by using pencil, paper, and hand calculators in a long-drawn-out process. These women became known as "computers."

The men who built the ENIAC chose six of these "computers" to program the giant machine. This involved setting dials and plugging a bewildering mass of black cables into the face of the computer, a different configuration for each problem:

> It was this job—"programming," they came to call it—to which just six of the young women were assigned: Marlyn (Wescoff) Meltzer, Ruth (Lichterman) Teitelbaum, Kay (McNulty) Antonelli and Frances (Bilas) Spence, as well as Ms. Bartik and Ms. Holberton. They had no user's guide. There were no operating systems or computer languages, just hardware and human logic. "The Eniac," says Ms. Bartik, now 71, "was a son of a bitch to program."[82]

After the female computers, in a public demonstration, programmed ENIAC to solve a trajectory problem at the official unveiling of the machine in February 1946, the engineers went out for a celebratory dinner. The programmers went home.

In the case of SAGE, Division 6 had initial responsibility for computer programming, but as the task grew in magnitude, Lin-

JAY FORRESTER (CENTER), ROBERT EVERETT (RIGHT, ARMS FOLDED), AND WHIRLWIND STAFF MEMBER STEPHEN DODD (SEATED) AT THE WHIRLWIND TEST CENTER IN THE BARTA BUILDING. RAMONA FERENZ IS SEATED AT THE DISPLAY. A NUMBER OF WOMEN SERVED AS TECHNICIANS AND SUPPORT STAFF FOR THE FORRESTER TEAM, BUT THEY ARE RARELY IDENTIFIED IN WHIRLWIND PHOTOGRAPHS OR NARRATIVES. *(Courtesy of MITRE Corporation)*

coln subcontracted with the RAND Corporation, a California nonprofit, Air Force–sponsored research center, to handle the bulk of the programming. Because programmers were virtually nonexistent, Division 6 and RAND recruited people from various occupations and backgrounds, training them for the exacting tasks. Specialists in music theory reportedly had a special aptitude for such work.

RAND established a systems development division in 1955 to deal with the programming for SAGE, but when this division's personnel numbered almost as many persons as all of RAND's, the division transmuted into the System Development Corporation (SDC), a nonprofit federal contract research center of the same kind as RAND. Provided with a $20 million contract for the task in 1957, SDC took responsibility for SAGE programming and for training the personnel to staff the combat and direction centers.[83]

After SDC and Lincoln engineers and scientists had become involved in SAGE programming, they faced the surprising reality that developing and deploying programs for large computers such as the AN/FSQ-7 demanded as sizable an investment of resources as the computer itself had had. Herbert D. Benington, a System Development Corporation scientist who took a leading role in developing the computer program for SAGE, recalls in similar vein:

> When we all began to work on SAGE, we believed our own myths about software—that one can do anything with software on a general-purpose computer; that software is easy to write, test, and maintain. . . . We had a lot to learn.[84]

They had so much to learn because a single programmer could not fully comprehend the complexity of a software program for a computer as large as the AN/FSQ-7 and a system as complex as SAGE. As a handful of SDC and SAGE programmers, led by MIT graduate electrical engineer C. Robert Wieser, delved more deeply into the SAGE task, they realized that they had to train thousands of programmers at a time when their total numbered in the hundreds. Those presiding over the massive endeavor found that the SAGE program would involve hundreds of thousands of lines of instruction code and that each line would cost about fifty dollars. By 1959, more than eight hundred people were working on the software for SAGE.[85]

Benington and Wieser and their associates realized that the programming project had to be organized rationally. They broke the project down into subprojects. These would develop into distinct modular-code packages that ultimately were combined into master programs suited for the distinctive geographical and weapon resources of each SAGE Direction Center.[86] With some justification, the programmers began to argue that software becomes the most complex component of a computer system. Those who presided over the programming of SAGE found it necessary to take a highly articulated organizational approach, one that resembled that being used by the systems engineers who were then conceptualizing, coordinating, scheduling, and monitoring the mammoth Atlas ICBM Project.

]]]]]] Systems Engineering

When we turn to the Air Force Atlas Intercontinental Ballistic Missile Project (Atlas ICBM), we find that by the mid-1950s a systems engineering organization was presiding over this project that involved thousands of contractors and subcontractors. By contrast, the SAGE Project still lacked coordinated management as late as 1957. The Air Force's Air Defense Command, Air Research and Development Command, and Air Materials Command had not harmoniously worked out a successful organizational and management scheme for SAGE. It had become a systems management problem seeking a solution. System builders in the 1950s were learning that the managerial problems of large-scale projects loomed as large as engineering ones.

In March 1956, with the installation of the first full-scale SAGE subsector pending, Forrester expressed misgivings about the lack of managerial leadership for the project. He called attention to the divided and limited managerial responsibilities of Lincoln Laboratory and the associated contractors. Lincoln had designed and developed the master computer; IBM presided over the manufacture of computers; Bendix Radio manufactured radar; Bell Telephone Laboratories provided the communications network; and Western Electric designed and constructed—and installed equipment in—direction and combat buildings. More than one hundred other industrial contractors contributed substantially to the project. But no Air Force, MIT, or industrial organization had the authority to preside over the coordination of such a large and heterogeneous assembly of contractors. Responsibilities were "spread . . . over all parts of the Air Force with no semblance of overall coordination except at the very top."[87]

The Air Force delegated some overall managerial responsibilities to Western Electric and the Bell Laboratories, but coordination problems remained unsolved. Forrester observed, "The situation is like putting up a large building with a collection of subcontractors but no general contractor responsible for the overall operation or the successful performance of the subcontractors."[88] Engineers familiar with the management of aerospace projects would have cited the need for a prime contractor or systems engineering organization. Forrester, however, did not wish his division of Lincoln Laboratory to take on the overall systems engineering role.

In June 1956, Forrester left the SAGE Project to become a professor of management at the recently established Sloan School of Management at MIT. Valley's promotion from head of Division 2 to the post of associate director of Lincoln may have been a factor influencing Forrester's decision, for confrontations between the strong-minded and strong-willed men might have become more frequent. Colleagues of Valley usually acknowledge his indispensable role in the establishment of Lincoln Laboratory and SAGE, but add that he was "difficult."[89] Valley's own anecdote about his aggressive behavior toward the Michigan air defense project leader bears out this contention. Valley found Forrester capable but insufferably arrogant, remarking after one encounter, "If I hadn't had to work with him, I'd have kicked his ass—with a steel point on my shoe, too."[90]

With Forrester gone, Division 6 managerial responsibilities fell on Everett's shoulders. Everett had been, as we have noted, a Forrester associate since the 1940s when they were graduate students at MIT, Forrester heading projects and Everett acting as an associate director. Everett expresses great admiration for Forrester's technical and managerial leadership, noting that he made insightful strategic decisions about research and development policy and that he, Everett, followed up on these by close attention to personnel and detail: "Boy, did I know what was going on."[91]

With an intuitive and insightful grasp of the overall nature of systems, Forrester would sometimes sharply criticize an approach taken by a staff member working on a system component. Because of Forrester's crisp, even cold, approach, he often left the erring staff member shattered. Everett picked up the pieces, reminding the distraught engineer that at least they could all take comfort in the fact that Forrester was usually right.

Before Forrester left Lincoln, Everett, like Forrester, did not want Lincoln to take over more of the systems engineering role for SAGE even though he saw negative consequences if another strong organization assumed that responsibility: "We would lose control of SAGE"; he feared that "we would be forced to work by indirection; we would be forced to watch while our creation was warped, perhaps beyond recognition, by strange hands" and "hard-headed supervisory" types brought in. Yet, he added, "I feel we should get out [of managerial responsibility] and pay the price."[92] He believed that, in compensation, Lincoln should be allowed to concentrate on research and development.

A year after Forrester had left, Everett changed his mind. In responding to an Air Force query as to whether or not Lincoln Laboratory would take on systems engineering responsibilities, Everett saw two alternative paths that Lincoln might follow: either to concentrate on research and development and leave systems management to another organization, or to take on systems management and demand the increased budget and staff this would require. He favored the latter course, as did the majority of the project managers. Reasoning that Division 6 had performed well because it not only developed the computer but also the related apparatus that integrated the computer into an information system, he proposed that Lincoln Laboratory should preside over the integration into a single air defense system of all of the major subsystems developed by other contractors, including the radar and communication networks.[93]

Because Division 6 had performed well as systems manager for the information-processing subsystem and because this constituted the core of the SAGE system, Everett proposed specifically that his division should act as systems engineer for the entire air defense system. He proposed a managerial structure similar to the one then used by the Air Force command responsible for the Atlas ICBM Project. He believed that Division 6 should function analogously as SAGE systems engineer for the Air Defense Systems Management Office established by the Air Force in 1957 at the Hanscom Air Base adjacent to Lincoln Laboratory. This office was intended to resolve the problem of divided responsibility among the various Air Force commands. Everett proposed that eventually Division 6 would relinquish the systems engineering role to a new systems engineering division at Lincoln, but that this new division would be heavily staffed by former Division 6 personnel.

According to Everett's plan, Division 6 would extend its purview beyond information processing to take responsibility for coordinating the design, development, and deployment not only of radar and communication links but of the intercepting weapons as well. He focused on the need to integrate weapons, such as aircraft, antiaircraft, and the BOMARC ground-to-air defense missile, into the system. At the time, a serious and critical "weapons integration" problem prevented direction centers from providing guidance and control to the intercepting weapons, both manned and unmanned. "The weapons contractors . . . [had] not . . . assumed responsibility for the successful integration of their weapons with the ground environment."[24]

Everett offered a definition of the amorphous systems engineering function. Division 6 would provide and continuously revise a "Master System Specification," he suggested, one pertaining to the design and performance of all subsystems and components. After verification by field testing, this specification would insure the compatibility of information processing, radar, communication, and weapons. The systems engineering organization would also determine the need for the development of new components for the system and would sign off on changes in existing components, making certain that adjustments would be made throughout the system to compensate for these changes. The interaction of operational plans and procedures and system design would also be superintended by the systems engineer. The information-processing subsystem would provide the information and control linkages for the heterogeneous components that constituted SAGE, transforming it into the organically coherent system envisioned years earlier by Valley's ADSEC committee.

• • •

]]]]]] MITRE

A variation of Everett's vision of Division 6 as a systems engineering organization came to pass through a drawn-out decision-making process involving the Air Force and MIT. The overall system coordination problem came to a head after Lincoln Laboratory, Bell Laboratories, and the Radio Corporation of America each declined the Air Force request for integration of the interceptor weapons into SAGE. The introduction of new interceptor aircraft and the Nike and BOMARC ground-to-air missiles complicated the task. The manufacturers would not accept the restriction that if they assumed the systems engineering responsibility they could not bid on SAGE hardware.[95]

The Air Force also had to face up to the past failures of several of its commands and offices to bring overall managerial control to the SAGE Project. The Air Defense Engineering Services Project Office, operated jointly by officers from the Air Material Command (AMC) and the Air Research and Development Command (ARDC); the SAGE Weapons Integration Group; the Air Defense Systems Management Office staffed by AMC, ARDC, and Air Defense Command officers; and the Air Defense Systems Integration Division—all had tried with limited success to coordinate the many organizations contributing to the SAGE Project.[96]

The Air Force had reason to doubt its ability to solve the systems management problems, and MIT did not favor Lincoln Laboratory's taking responsibility for managing the SAGE industrial contractors. With Lincoln Laboratory or another division of MIT in such a role, academia-industry tensions could develop with companies that hired MIT graduates and gave research and endowment funds to MIT. Earlier, the MIT administration had resisted Lincoln Laboratory's expansion into computer programming and computer manufacture, in part because this would have moved the institute even further from its primary mission of education, placing it in competition with the industrial sector.

Pressured by the Air Force to find a solution to the problem outside of Air Force commands and Lincoln, Killian and Stratton agreed to help the secretary of the air force find or found an organization to take over systems engineering. They considered the possibility of establishing a nonprofit organization similar to the System Development Corporation. Killian, Stratton, and Douglas turned to James McCormack, who had recently retired from the Air Force as a major general and become the MIT vice president in charge of relations with the military.

McCormack merits our attention as a prominent figure in the military university-industrial network. A West Point Military Academy graduate of 1932 and a Rhodes Scholar at Oxford University where he received a degree in languages, McCormack also had a master's degree in civil engineering from MIT. He served in the Army Corps of Engineers, becoming chief of the Construction Branch of the War Department General Staff in 1943. After the war, he became a member of several War Department policy groups, then transferred to the Air Force in 1950. He rose rapidly in the newly formed Air Force Air Research and Development Command, becoming vice commander in 1952, a command with responsibilities for SAGE, the Atlas ICBM, and other major Air Force projects.

McCormack, who strikingly resembled the duke of Windsor, greatly impressed Lincoln Laboratory and MIT people not only because of his achievements but also because of a remarkable ability to summarize, without reference to notes, the gist of numerous, complicated luncheon meetings, despite his having consumed an impressive number of martinis. These seemed only to stimulate his powers of memory and analysis.[97] His intelligence, influence, and style suited him well for the Cold War MIT and Cambridge scene.

McCormack and others involved in the decision making came down to a choice between the System Development Corporation, which wanted the assignment, or the establishment of a new organization with heavy representation of personnel from Division 6.[98] Finally, they opted for the latter, organizing a new not-for-profit Federal Contract Research Center that took the name MITRE Corporation.[99] It resembled the RAND Corporation as well as the System Development Corporation. MITRE as systems engineer would report to an upgraded Air Defense Systems Management Office

(ADSMO) under the command of General Bernard A. Schriever, who also headed the Atlas ICBM Project. Schriever had resorted to using a systems engineering organization earlier when he gave the Ramo-Wooldridge organization managerial responsibilities for the Atlas Project.

Clair W. "Hap" Halligan, director of military engineering at Bell Laboratories, who had often served as a consultant to Lincoln, became MITRE president. Halligan's Bell Laboratories experience and his judiciousness impressed both the director of Lincoln Laboratory as well as McCormack, who nominated Halligan for the position. Everett moved voluntarily from Division 6 to become technical director; John Jacobs, also a principal engineer in Division 6, became associate technical director. About 300 people, most of Division 6, transferred to MITRE to provide the start-up staff. Through a subcontract from MIT, the Air Force appropriated $13 million in start-up funds for MITRE.

MITRE's systems engineering responsibilities for the SAGE defense system did not develop as McCormack and its creators intended. The Air Force did not vest in MITRE the managerial authority needed to effectively direct and coordinate the various SAGE contractors. MITRE managers and engineers had to be satisfied with monitoring deployment, calling contractors' attention to emerging problems of coordination and integration, suggesting solutions, and attempting to persuade contractors to respond in appropriate and timely fashion. We shall discover Ramo-Wooldridge exercising more control over the Atlas contractors than did MITRE over SAGE contractors.

]]]]]] Learning Experience

Critics have called SAGE the best peacetime air defense system ever deployed. Behind this quip lies the fact that SAGE was designed at a time when long-range bombers loomed as a threat, and that it was then deployed when long-range missiles had become an increasing threat. In 1955, the Technological Capabilities Panel headed by Killian advised President Eisenhower that existing and planned air

AIR FORCE ENLISTED MEN MANNING TERMINALS IN A DIRECTION CENTER. *(Courtesy of MITRE Corporation)*

defenses could not prevent nuclear-armed bombers—much less missiles—from inflicting unacceptable destruction in U.S. cities.[100]

Despite rising doubts about SAGE's effectiveness, the Air Force organized a dramatic public display of the inauguration of the first operational SAGE sector. On 7 August 1958, in Kingston, New York, an engineer activated a computer and in a few moments a BOMARC missile rose from Cape Canaveral, Florida, guided by a SAGE computer to intercept a simulated enemy bomber.[101] In the following months, twenty-two of the planned thirty-two direction centers were built; those in the less vulnerable Midwest were never built. In the 1980s, the Air Force shut down all of the SAGE direction centers. Robert Everett, head of MITRE, recalls that the Air Force made no clear-cut decision to abandon SAGE but "just let it fade away."[102]

Can SAGE be labeled simply a failure? At the bar of history, critics will emphasize SAGE's inadequacies as a defense against both bombers and missiles. Supporters, however, will stress that it became a learning experience of surpassing influence. Facing the winding down of SAGE, MITRE, for instance, sought ways to use the know-how acquired by its engineers and technicians who had nurtured SAGE. Happily for them, their learning experiences in computers and systems engineering transferred well into military-funded systems.[103] The command, control, and communications system that interconnected SAGE facilities, for instance, became the model for the Air Force's massive system for connecting far-flung bomber bases and headquarters into a single network. Subsequent to SAGE, the Air Force funded development of similar computerized command, control, and communications systems including those for the Strategic Air Command Control System, the North Atlantic Air Defense Command, the NATO Air Defense Ground Environment, and the World Wide Military Command and Control System.[104] Engineer Jacobs, a Division 6 and MITRE veteran, characterizes SAGE "as a rich source of experience" that "has been used as the example of how to and how not to design and evaluate systems."[105]

Much like the Erie Canal project early in the nineteenth century became in effect the leading American engineering school of its day, the SAGE Project became a center of learning for computer engineers, scientists, and technicians. Charles Zraket, who like many of his MITRE colleagues, studied at MIT, worked at Lincoln Division 6, and then transferred to MITRE, has summarized the SAGE learning legacy as "profound."[106] Later MITRE technical director, vice president, and chief operating officer, Zraket recalls that the SAGE Project afforded him and others like him invaluable experience with computer time-sharing; with designing real-time computer displays; and with various software-programming techniques.

SAGE became a "tremendous technology transfer" agent.[107] After the SAGE experience, its engineers and scientists took leading roles in establishing MIT's Project MAC, which introduced computer time-sharing; started a number of university computer science departments; helped create the ARPANET, predecessor to the Internet; and established companies like the Digital Equipment Corporation (DEC).[108] At least forty-six of the engineers who worked on the development of the Whirlwind computer moved to industry, where they transferred "vital Whirlwind know-how."[109] Everett compares the learning experience to a voyage of discovery: "I have a picture in my mind of Columbus and his crew discovering a new world. . . . Jay [Forrester] was our Columbus and so we discovered strange and wonderful things."[110]

Not only SAGE people but organizations associated with SAGE acted as technology transfer agents. The System Development Corporation trained hundreds of programmers for SAGE and continued to do so for other projects.[111] The MIT Lincoln Laboratory has remained a major mission-oriented research center. MITRE continues to develop command and control systems as well as other advanced technology. The SAGE legacy recalls the prophetic words of Ridenour, head of the Air Force Scientific Advisory Board, who told Killian in 1950 that MIT should undertake the laboratory proposed by the Valley committee not only because of a national security need but also because the laboratory would insure MIT's leadership role in electronics research.

III Managing a Military-Industrial Complex:

Atlas

Systems engineering is the design of the whole as distinct from the design of the parts. Systems engineering is inherently interdisciplinary because its function is to integrate the specialized separate pieces of a complex of apparatus and people—the system—into a harmonious ensemble that optimally achieves the desired end.

Simon Ramo

. . . "tomorrow's man" with a big project concept . . . often runs his command post in a grey flannel suit or tweed sports coat and slacks, [and] . . . decorates his command post with an impressionistic oil painting of the U.S.'s first liquid-fuel rocket superimposed upon a plumed Chinese war rocket supposedly used by the Kin Tartars at the siege of Kaifeng . . .

***Time* (1957)**

▼ "Tomorrow's man," General Bernard Schriever, and Simon Ramo presided over dramatic changes in the management of large aerospace projects. They introduced a refined systems approach that facilitated the management of great complexity. They designed a management structure that displaced the established airframe manufacturers as prime contractors of aerospace projects. They nurtured a culture attractive to high-technology engineers and scientists, who, in turn, provided the talent needed to manage and to solve the difficult research and development problems of Atlas, the country's first successful intercontinental ballistic missile project.

Begun in 1954, the Atlas Project became the first of several projects that constituted an ICBM program culminating in the production of three missiles, the Atlas, the Titan, and the Minuteman. By 1957, the program involved 17 principal contractors, 200 subcontractors, and a workforce of 70,000. A year later, the program launched Atlas and Titan test missiles that proved their feasibility with an accurate 5,000-mile flight.[1] Initiated in 1958 and deployed in 1962, the Minuteman ICBM superseded Atlas/Titan, but this chapter concentrates on the Atlas project as representative of the management of research and development in the first postwar wave of major military-industrial projects.

Earlier engineering projects, such as hydroelectric dam-building, though formidably large, had not required nearly as much research and development. Companies building chemical process plants had encountered complex research and development problems but not on the scale of the Atlas Project. Creating Atlas became, like the SAGE Project, a major learning experience for the managers, engineers, and scientists involved.

Whereas the learning in the case of SAGE was mostly about computers, in the case of the Atlas Project most of the learning was about managing large-scale projects. Systems engineering as now practiced on large-scale civil as well as military projects had its substantial and sustainable beginnings during the Atlas Project. We shall follow, then, events and trends culminating in substantial changes in deeply embedded and long-standing managerial policies of military organizations and industrial corporations that, like supertankers, find making course changes agonizingly slow because of their encountering extremely high momentum.

]]]]]] Agents for Change

After World War II, the leadership of the U.S. Army Air Force insisted that fighter and bomber aircraft had played a leading role—if not *the* leading role—in the allied victory. Air Force generals had launched their careers by flying bombers and fighters; enlisted men had become expert in maintaining the aircraft; Air Force commands had established close alliances with aircraft manufacturers. Together they polished a routine involving Air Force funding, oversight management, and industrial corporations acting as prime contractors and subcontractors for the design-to-deployment functions involved in the production of aircraft.

Only an exceptional Air Force general would have looked for a new weapon to supplement, even displace, manned aircraft. Henry "Hap" Arnold, a five-star general and chief of staff for the Army Air Force, was one of these. He called for the development of long-range missiles with greater range, accuracy, and destructiveness than the V-2 missiles the Germans had used at the end of the war.[2] The farsighted Arnold argued that missile development would be impossible without the involvement of the country's leading engineers and scientists. Well informed about research and development, Arnold took into account the part played by Germany's engineers and scientists in the development of the V-2 missile.

Arnold's commitment to military-funded research and development had gathered strength during the years when he was stationed in California. He had a prominent place within the network of people and institutions nurturing the aircraft—and later the aerospace—industry that grew up in the Los Angeles region between the two world wars and afterward.[3] From 1931 to 1936, Arnold commanded the Air Force base at March Field near San Bernardino, California, and about fifty miles from the California Institute of Technology, where he became close friends with Professor Robert Millikan, Nobel laureate physicist and head of Cal Tech's governing executive committee.

Seated at Millikan's kitchen table, they often talked half the night, discussing the engineering, economics, and politics of the aircraft industry. Through Millikan, Arnold met Theodor von Kármán, a leading European aerodynamicist who in 1930 had become professor of aeronautics and head of the Cal Tech Guggenheim Aeronautics Laboratory (GALCIT), a laboratory that gave Cal Tech a preeminent role in the development of the science of aerodynamics. Arnold helped bring research and development contracts to von Kármán, including those funding the study in 1939 of rocket propulsion, especially rocket-assisted takeoff for aircraft. These studies bore fruit with the establishment by von Kármán and his

Cal Tech associate Frank Malina of both the Aerojet Engineering Company, later Aerojet-General (the "General Motors of Rocketry"), and the Army-funded Cal Tech Jet Propulsion Laboratory in 1944. Not all scientists had the foresight of von Kármán: Vannevar Bush, influential MIT engineering professor, dean, and government policy adviser, remarked, "I don't understand how a serious scientist can play around with rockets."[4]

Arnold composed a memorandum in 1944 contending that the outcome of future wars depended more on new missiles than on pilots.[5] He then asked von Kármán to assemble a small panel of scientific and engineering experts to prepare a study of the future of air warfare, adding, "The security of the United States of America will continue to rest in part on developments instituted by our educational and professional scientists."[6] Von Kármán recalls that the general said to the panel:

> I see a manless Air Force. . . . For twenty years the Air Force was built around pilots, pilots, and more pilots. . . . The next Air Force is going to be built around scientists—around mechanically minded fellows.[7]

The appointed panel surveyed wartime achievements in aeronautics and rocketry, visited defeated Germany, and presented to Arnold in July 1945 a secret report, from which emerged the influential and long-lived Air Force Scientific Advisory Board that we last encountered calling for the overhauling of air defense.

The Von Kármán report singles out jet propulsion and jet-powered aircraft, not ballistic missiles, as the key future technologies, von Kármán's retrospective claim that he was among the early missile proponents notwithstanding. We should note, however, that the call for jet-powered aircraft was a forward-looking proposal at that time, when bombers were propeller driven. Von Kármán's few references to long-range missiles envisioned air-breathing pilotless aircraft (cruise missiles).[8]

Before he retired in 1947, Arnold, with his close friend Donald Douglas, head of Douglas Aircraft, facilitated the establishment of the RAND Corporation, a nonprofit research center that would advise the Air Force on strategic policy in an era of long-range missiles and nuclear weapons and that would become involved in providing programmers for SAGE. When we turn to the spread of the systems approach in the 1960s, we shall observe the central role of RAND scientists, social scientists, and engineers.

]]]]]] Convair's Atlas, 1946–1950

Arnold and other military and science policy makers who were impressed by the potential of long-range ballistic and other missiles in the 1940s supported related research and development on a modest scale. Funded by military contracts amounting to about $50 million by the end of the war, several companies—notably North American Aviation, Bell Aircraft, General Electric, and Consolidated Vultee—maintained small missile divisions formed during the war. In addition, the Glenn L. Martin Company, Curtiss-Wright, Republic Aviation, and Northrop Aircraft had small missile projects. Unlike the Army Ordnance Department, the Air Force did not rely on a team of émigré German rocket scientists; instead, it invested in American academic scientists and the aircraft industry.

As the number of missile projects increased to about fifty—with the Army Air Force sponsoring half of these—the National Academy of Sciences set up the Technical Evaluation Group to survey the field and evaluate the quality of the various efforts. Despite finding the quality of a number of the missile projects low, Professor Francis Clauser of Cal Tech, a member of the evaluation group, concluded that thousands of American engineers were learning about missiles—especially guidance of missiles—from working on the projects.[9] Nevertheless, by March 1947 the military services had reduced the number of projects to nineteen.

Concerned about overlapping missile projects sponsored by Army Ordnance, the Army Air Force, the Navy Bureau of Ordnance, and the Navy Bureau of Aeronautics, the newly created Department of Defense attempted after 1947 to define the respective responsibilities of the services for missile research and development. As a result, by 1950 it had become reasonably clear that the Air Force, recently separated from the Army, had responsibility for long-range surface-to-surface missiles.[10] Which service should take charge of the development of intermediate-range surface-to-surface ballistic missiles remained unclear.

In January 1946, the Air Force awarded Consolidated Vultee Aircraft Corporation (Convair) a $1.4 million contract (MX-774) to support a year's study of alternative designs for a 5,000-mile intercontinental ballistic missile for delivering a 5,000-pound warhead within 5,000 feet of the target. Within several years, the MX-774 metamorphosed into the Atlas Project, a transition shortly to be described. In 1946, however, the Convair engineers proposed to do only conceptual designs for a subsonic, jet-powered missile and for a rocket-propelled ballistic missile. ("Ballistic" refers to a missile that follows a gravity-shaped trajectory after initial propulsion, a flight similar to the one taken by a projectile from a long-range gun.)[11]

After producing many airplanes during the war, Convair was suffering from the postwar drop in aircraft procurement and thus welcomed the study contract, despite its relatively small size as compared to aircraft contracts. Karel J. "Charlie" Bossart, a modest, tactful engineer whom his coworkers considered one of the finest in the country, took charge of the study project. One colleague characterized Bossart as "one those rare types who make practical engineering out of very complex theories."[12] Another fellow engineer describes Bossart as "one of the kindest, most good-natured persons I have ever known," who let others believe they were partners in inventing a dramatically innovative missile design. Belgian-born, Bossart had graduated with a degree in mining engineering from the University of Brussels in 1925, then won a scholarship from the Belgian-American Education Foundation to study at MIT, where he specialized in aeronautical structures.

The Convair team reduced missile weight in several ingenious ways. Since the V-2 had heavy interior fuel tanks, Bossart used the missile structure as the fuel tank, similar to the way airplanes use wings to carry fuel. The missile becomes a flying fuel tank with a power plant at the rear and instruments at the front.[13] To further reduce weight, the engineers removed the interior stiffening braces of the V-2, using pressurized nitrogen gas to sustain the dime-thin walls of the "steel balloon."[14] The pressurized nitrogen also forced the cryogenic fuels, liquid hydrogen and oxygen housed in separate compartments, into the combustion chamber.

Although the entire V-2 missile made the flight from launch to target, this required that the missile be shielded from the destructive heat generated by reentry through the atmosphere. Bossart and team designed their missile so that a nose cone carrying the warhead separated from the body of the missile after the rocket engines accelerated the missile to the desired velocity. Only a heat-shielded nose cone, or reentry vehicle, loaded with explosives would fly to target. This design greatly reduced the weight of the missile and the rocket engine-thrust requirement.

The Bossart team also introduced swiveling rocket engines to direct the flight of the missile during the propulsion stage. By contrast, V-2 had used vanes in the exhaust stream, which reduced the effective thrust of the rocket engines.[15] For guidance, the Convair engineers chose an onboard, gyrostabilized autopilot capable of receiving corrections from ground radio, a guidance system later used in testing full-scale Atlas missiles.[16]

Obtaining at engine cutoff the precise missile velocity necessary for propelling the nose cone to its distant target also presented a critical problem. Realizing the difficulty of adjusting engine thrust so delicately, the engineer responsible for guidance solved the problem by adding, near the rear of the airframe, two small vernier rockets whose thrust would provide feather-touch control.[17]

Increasing Cold War tension frustrated continuing development of the design of the Convair missile. Giving funding priority to the production of long-range bombers ready for combat, the Air Force did not appropriate additional funds for the Convair project. The Air Force first suggested to Convair that it stretch its funds appropriated for 1946 into 1948. Then, in July 1947, the Air Force terminated the MX-774 Project because influential staff officers assumed that ballistic missiles did not promise "any tangible results in the next 8 to 10 years."[18]

As a consequence, the Bossart team had sufficient funds for testing only three—not ten, as originally planned—small, 1,200-pound, 8,000-pound-thrust test missiles. (The weight of a full-scale missile with nuclear warhead was projected as a quarter of a million pounds.) After successful static tests in the company's facilities near San Diego, California, the first of the test missiles arrived at White Sands Proving Grounds, New Mexico, for flight testing in July 1948. After a short rise of less than a minute, the missile tumbled, plunged to earth, and exploded with more than half of its fuel on board. In September, missile number two performed little better. In both failures, the propulsion system prematurely burned out. During December, the third missile flew for fifty-one seconds before it exploded.[19] Though they are discouraging, failures are expected by experienced engineers testing highly innovative designs.

Without Air Force funds to continue testing and design, Convair invested its own, assuming that the Air Force would take up the long-range missile project again. On the negative side, however, Bossart noted that the Air Force provided no specific engineering advice during the design process and that low-ranking Air Force officers, whose visits to the Convair plant could be "counted on the fingers of one hand," monitored the project.[20] Furthermore, Air Force policy then favored the Navaho missile project of North American Aviation and the Snark project conducted by Northrop Aircraft simultaneously with the MX-774. Both the long-range Navaho and Snark were aerodynamic (cruise) and jet-driven missiles that appealed to many officers because they resembled traditional airplanes—without pilots.[21]

]]]]]] **Conservative Momentum**

Despite Convair's internal funding of the MX-774 missile project, renamed Atlas, several years passed before the Air Force took another look at the project. In 1947, an Air Force Staff review panel, in analyzing the future of missiles, predicted that "for the next ten years, at least, the subsonic bomber will be the only means available for the delivery of long-range (1,000 miles and over) air bombardment."[22] The panel assigned the development of a long-range surface-to-surface missile a priority ranking below air-to-air, air-to-surface, short-range surface-to-surface (150 miles), and surface-to-air missiles. In 1947, President Harry Truman's high-level Air Policy Commission reinforced Air Force policy. In 1949, the National Security Council advised the president that the highest national priority should be given to the production of atom bombs and B-36 bombers. The council did not mention long-range missiles.[23]

A conservative momentum, or inertia, frustrated a move to long-range missiles. The National Security Council was taking into account the existing technology and organizational support dedicated to long-range bombers and the national security–endangering disruptions and delays that would follow upon radical change. The Air Force was dominated by officers in peril of becoming deskilled if missiles displaced bombers. Robert Perry, an experienced analyst of military policy, argues that officers who flew manned bombers before and during World War II had an affection for them. They regarded the flying of airplanes as a way of life, the essence of a military culture. They dismissed as misfits Air Force officers who were engineers and scientists first, pilots only second.[24] In similar vein, Edmund Beard, author of *Developing the ICBM* (1976), suggests that the Air Force bureaucracy instinctively resisted radical innovation as disrupting established procedures and diluting the power of the bureaucracy. He estimates that institutional inertia delayed the initiation of a sustained long-range ballistic missile project by at least six years.[25]

This inertia, or conservative technological momentum, consists of both institutions and hardware. Not only do human values and routines anchor organizations, but physical objects with specific characteristics generate a resistant mass weighted in favor of the status quo. Elting Morison, a cultural historian of technology, has delineated the force of this conservative momentum.[26] He tells of the effort, at the turn of the century, of a young naval officer, William S. Simms, to introduce a demonstrably improved method of aiming naval guns at sea. The resistance of an entrenched naval bureaucracy to a junior officer's innovation that would have greatly increased the Navy's effectiveness exasperated Simms. Only when President Theodore Roosevelt intervened was the institutional inertia overcome and Simms able to institute his reform. We shall soon encounter civilian authorities who play similar roles in overcoming internal resistance within the Air Force to a large-scale intercontinental ballistic missile project.[27]

Ernest Schwiebert, an Air Force historian, commenting on the resistance to ballistic missiles, writes:

> The hurdle which had to be annihilated in correcting this misunderstanding was not a sound barrier, or a thermal barrier, but rather a mental barrier, which is really the only type that man is ever confronted with anyway.[28]

We should not conclude, however, that only entrenched Air Force officers displayed mental barriers. Bush, whose enormous prestige as a presidential science adviser stemmed from his leadership of the Office of Scientific Research and Development during World War II, stated in testimony before a Senate committee in 1945:

> There has been a great deal said about a 3,000 mile high-angle rocket.
>
> In my opinion such a thing is impossible and will be impossible for many years. The people who have been writing these things that annoy me have been talking about a 3,000 mile high-angle rocket shot from one continent to another carrying an atomic bomb, and so directed as to be a precise weapon which would land on a certain target such as this city.
>
> I say technically I don't think anybody in the world knows how to do such a thing, and I feel confident it will not be done for a very long time to come. I think we can leave that out of our

thinking. I wish the American people would leave that out of their thinking.[29]

Scientists and engineers who doubted the feasibility of developing a long-range ballistic missile in the near future usually cited high atomic-warhead weight (5,000 to 10,000 pounds) as a major hurdle to be overcome.[30] They also questioned the possibility of designing a missile guidance system that would insure reasonable missile accuracy. (The V-2 usually missed its target by about 10 miles at a 200-mile range.) Experts felt that finding a design and materials for a heat shield to protect the nose cone from the enormous temperatures encountered during atmospheric reentry also posed a critical problem.

Advocates of ballistic missiles within and outside the Air Force countered that generously funded research and development could bring early solutions to the problems. They contended that Air Force policy makers customarily exaggerated the severity of problems, then failed to allocate funds to solve them.

)))))) Confluence for Change

Chance, contingency, and confluence eventually broke the conservative momentum and made possible the initiation of a full-scale Atlas Project. Radical change, however, took an indirect, long, and tortuous path. It began with stirrings in the media. In 1950, several newspaper columnists, including the highly regarded Hanson Baldwin, military expert of the *New York Times,* drawing upon his excellent sources in the Defense Department, wrote:

> It is true that the finished guided missile of great accuracy and an almost human brain is still in the future. . . . But it is clear that these problems are being licked. Russia's progress in the guided missiles field is an unknown quantity in this country, but it seems probable that she is ahead of us. . . .[31]

About the same time, Democratic Senator Lyndon Johnson, a

member of the Armed Services Committee, publicly warned that the Russians might be two years ahead of the United States in guided-missile development. President Truman responded by naming K. T. Keller, president of the Chrysler Corporation, as a special adviser to the Defense Department to coordinate guided-missile work and—according to some reports—to establish a "Manhattan-type" missile project. Keller, however, whose staff never numbered more than ten, disavowed having been given such a lofty charge, contenting himself with a modest effort to eliminate duplication of effort among the services. His appointment seems to have been in part a public relations effort conceived in response to press criticism of administration inactivity.

Despite this flurry of publicity and presidential action, the Air Force remained unmoved.[32] In late 1950 and early 1951, its funding of two studies of long-range missile feasibility, one by the RAND Corporation and the other by Convair, can also be seen as a defensive maneuver—staking out a claim without investing heavily in the development of the property. The requirements laid down for the Convair study—as contrasted with a design and development project—specified a heavy, 8,000-pound atomic warhead, a range of 5,000 miles, and an unrealistic circular probable error (radius of a circle within which half of all the warheads will land) of 1,500 feet.[33] The Air Force continued to define large problems while at the same time appropriating small funds.

In 1952 and 1953, a confluence of scientific and technological events substantially altered Air Force policy. In 1952, U.S. nuclear tests proved the feasibility of a light nuclear warhead of enormous destructiveness and the Atomic Energy Commission founded the Lawrence Livermore Laboratory in Berkeley, California, to accelerate the development of thermonuclear weapons. The explosion of a fusion device by the Soviet Union in the following year as well as U.S. intelligence reports about Soviet long-range missile projects greatly increased the determination of missile advocates in the United States to initiate a large-scale ICBM-development project. The recently inaugurated President Eisenhower placed high-ranking civilians who supported missile development in the Defense Department.

In the spring of 1953, Edward Teller and John von Neumann, scientists intimately acquainted with and supportive of the development of thermonuclear weapons, informed the Air Force Scientific Advisory Board of the feasibility of substantially decreasing the weight of a thermonuclear warhead. They projected one that would weigh only about a ton and would be capable of delivering the explosive power of one to two megatons of TNT. (Earlier estimates had the weight of the warhead at about four tons.) Both Teller and von Neumann wielded enormous influence among Air Force officers.[34]

Stimulated by the predictions of Teller and von Neumann, the Air Force board set up a panel headed by von Neumann, which included Teller, to consider the effect of the lighter warhead upon Air Force weapons development, including airplanes, cruise missiles, and ballistic missiles. Other leading scientist members included Norris E. Bradbury, director of the Los Alamos Laboratory; Hans Bethe, a physicist who advised the military about nuclear weapons; and Herbert York, director of the recently formed Livermore Laboratory from which the Air Force expected the new generation of hydrogen bombs to emerge.

Even though the military's investment in ICBM research and development was notably small as compared to that in long-range bombers, the panel recommended in June 1953 a greater emphasis on missile development. This encouraged Trevor Gardner, recently appointed Air Force assistant for research and development, to form a Strategic Missiles Evaluation Committee—soon thereafter called the Teapot Committee[35]—which would weigh the implications of lighter thermonuclear weapons for ICBM development.

The Teapot Committee evaluated the credibility of troubling intelligence information that suggested an increase in long-range missile activity in the Soviet Union. This information might have come from radar installations in Turkey as well as from reports of missile accidents and rocket-engine development in the Soviet

Union.[36] Reports of the decline in Russian bomber production also pointed to an emphasis on intercontinental as well as intermediate-range missiles. This decline led some American analysts to the alarming conclusion that while the United States was investing enormously in long-range bombers at the expense of developing missiles, the Soviet Union was avoiding what could be considered a move in the wrong direction. (Several years later, intelligence suggested that, in fact, the Russian emphasis had been on construction of intermediate-range missiles based in Eastern Europe and aimed at targets in Western Europe.)[37]

In February 1954, the committee reported prudently and cautiously that available intelligence provided evidence that the Soviet Union appreciated the importance of intercontinental ballistic missiles and that the country was involved in some activities whose objective might be the development of an intercontinental ballistic missile. The committee did not rule out the possibility that "the Russians are ahead of us." Experienced committee members, accustomed to reading between the lines of policy reports, wanted a statement that would not "frighten" the casual reader and, at the same time, would alert the "proper reader" and accomplish "all that we should seek to accomplish with the available facts." In an early draft of the February report, Simon Ramo, who prepared the final report for the committee, argued in a more radical vein that most members of the committee believed on the basis of available evidence that "the Russians are probably significantly ahead of us in long-range ballistic missiles."[38]

]]]]]] Trevor Gardner Moves the Air Force

With intelligence reports causing concern among "proper readers," with decreased warhead weight breeding optimism among designing engineers, and with Livermore Laboratory promising to deliver lightweight thermonuclear warheads, Eisenhower appointees at the Department of Defense initiated a train of events that culminated in the formation of the full-scale Atlas Project within the context of an intercontinental ballistic missile program. This was the largest

and most costly military-funded research and development program in history—the Manhattan Project notwithstanding.[39] A
crisis mentality, justified or not, helped break the inertia of bureaucracy.

Charles Wilson, former head of the General Motors Corporation, whom Eisenhower appointed secretary of defense, decided
that the country was ineffectively spending money on overlapping
and unproductive missile development programs. His air force secretary, Harold Talbott, accordingly assigned to Trevor Gardner, Air
Force assistant for research and development, the mission of evaluating the Air Force's strategic missile efforts. Gardner determinedly set about defining a program, establishing organizations,
and securing funds to create a fleet of operational, intercontinental
ballistic missiles.

Born in Cardiff, Wales, trained in engineering and business
administration at the University of Southern California, Gardner
worked for about a year at the General Electric Company and subsequently taught industrial management at Cal Tech. During
World War II, he headed engineering development projects (a
rocket project among them) at Cal Tech for the Office of Scientific
Research and Development. After the war and after the General
Tire and Rubber Company bought Aerojet-General, the company
that von Kármán and colleagues at Cal Tech had formed to build
rocket engines, Gardner became general manager and executive
vice president of General Tire and Rubber. In 1948, he left the
company to establish and become vice president and later president
of Hycon Engineering Company, a small electronics firm, located
near Cal Tech. From Hycon, he went to Washington to serve as
an Air Force special assistant, taking with him experience and credentials in the aerospace and electronics advanced technology community.

Associates described Gardner as sharp, grating, abrupt, irasci-
ble, cold, unpleasant, and a bastard.[40] Ramo, who was at General
Electric at the same time as Gardner, remembers him as an interesting, "rebellious, very bright guy," a person of high integrity but
not gifted as an engineer.[41] Some of the negative reactions to Gard-

ner came from Air Force generals who resented a civilian giving orders contrary to their go-slow policy on missiles. Other officers feared that Gardner wanted to become czar of a new missile empire similar to the Manhattan Project, which would have removed the missile program from Air Force jurisdiction.[42] General Bernard Schriever, who worked closely with Gardner on a number of occasions, recalls that he "could be real nasty. . . . He did not tolerate stupidity," but that he acted effectively as a catalyst and mover for the ICBM program.[43] In a 1963 obituary editorial, the *New York Times* called him a "brilliant man of wide horizons," commending him not only for serving as a prime architect of the ballistic-missile program but later as one of the chief artisans of the Arms Control and Disarmament Agency, which he served as a member of its General Advisory Committee.[44]

Donald Quarles, the assistant secretary of defense for research and development in the Eisenhower administration, also took an active part in defining and promoting the Atlas Project. His background, like Gardner's, led him to place confidence in firms experienced in electronics and systems engineering. He had served as director of, first, the transmission development department, then of apparatus development for Western Electric, the manufacturer of electrical and electronic equipment for the Bell Telephone system, before becoming vice president of Western Electric in 1948. In 1952, Quarles, whom Schriever describes as solid, stable, and conservative in approach, became president of the Sandia Laboratories, an Atomic Energy Commission–funded laboratory in Albuquerque, New Mexico, operated by Western Electric and responsible for nuclear weapons applications.

]]]]]] The Teapot Committee

Gardner markedly accelerated missile development by establishing, as we have noted, the Air Force Strategic Missiles Evaluation Committee, or Teapot Committee. For advice in organizing the committee, Gardner had turned to Ramo, another Cal Tech graduate whose management of Air Force projects at the Hughes Aircraft Company favorably impressed him as much as the slack, uncoordinated management elsewhere disturbed him. Believing that John von Neumann's immense reputation—some compared him as a mathematician to Albert Einstein as a physicist—would attract science and engineering leaders to the committee, Ramo recommended von Neumann for committee membership. Von Neumann also seemed a wise choice because he had not been drawn into the J. Robert Oppenheimer controversy that had polarized the science community after the withdrawal of his security clearance in 1953.

Von Neumann came to the United States before World War II to become a lecturer at Princeton University and a scholar at the Princeton Institute of Advanced Study. Associates found him amiable, considerate, and mature as well as cerebrally dazzling. Not only did he gain recognition for his theoretical mathematical papers but also for those on applied mathematics, including ones on hydrodynamics and aerodynamics that later helped him in advising on ballistic-missile development. His experience in solving complicated mathematical problems on digital computers, such as those encountered when projecting a missile's trajectory, also recommended him for service on the Teapot Committee. After doing applied mathematics for the military during World War II, he became deeply involved in advising the military and the Atomic Energy Commission. Serving on both the AEC General Advisory Committee and the Air Force Science Advisory Board, von Neumann acted as a gatekeeper between those developing the missile system and those designing its warhead.[45]

Von Neumann's deep commitment to military consulting stemmed in part from his concern about the spread of Soviet authoritarian socialism, an ideological stance similar to Teller's. In 1951, von Neumann wrote to Lewis Strauss, head of the Atomic Energy Commission, "I think that the US-USSR conflict will very probably lead to an armed 'total collision' and that a maximum rate of armament is, therefore, imperative."[46] Von Neumann liked and got along well with high-ranking military officers and they with him.

Schriever found von Neumann's ability to anticipate questions remarkable and that if von Neumann "liked you, he became approachable, warm, and friendly."[47] Herbert York testified that "other smart people commonly said that John von Neumann was the smartest person they had ever known."[48] York had in mind von Neumann's mathematical analysis of quantum theory, his original work in economic theory, his development of game theory used by scientists and social scientists, and his invention of a computer architecture widely adopted by early mainframe designers. Despite his enormous learning and prestige, von Neumann took pains not to dominate committee discussions, playing instead a Socratic role, often asking of fellow committee members seemingly naive questions with such remarks as "Forgive me, I do not understand this," and in so doing bringing them to clarify their own concepts.[49]

For the Teapot Committee, Gardner assembled an elite group with strong representation from Cal Tech and the West Coast. We should recall that the committee associated with the SAGE air defense system had strong representation from MIT and the East Coast. The membership of the Gardner committee included Charles C. Lauritsen, a Danish-born Cal Tech physicist; Clark B. Millikan, a Cal Tech aeronautical scientist who headed the Cal Tech Guggenheim Laboratory; Louis Dunn, director of Cal Tech's Jet Propulsion Laboratory; Hendrik Bode of the Bell Telephone Laboratories, author of an influential book on synergy and technical integration; Lawrence A. Hyland of Bendix Aviation Corporation; George B. Kistiakowsky, a Russian-born chemist from Harvard University, later presidential science adviser to Eisenhower; Allen E. Puckett of California's Hughes Aircraft Corporation; and Jerome Wiesner, MIT electrical engineer, later the science adviser to President Kennedy, and then president of MIT.

Simon Ramo and Dean Wooldridge, both Cal Tech graduates, recently of Hughes Aircraft, and newly the founders of the Ramo-Wooldridge Corporation, also served on the committee. They and their company acted, in addition, as the committee technical and administrative staff, with responsibility for organizing and presenting background studies and arranging briefings. In addition to

DEAN WOOLDRIDGE (LEFT) AND SIMON RAMO ON *TIME* MAGAZINE COVER, APRIL 29, 1957. *(Courtesy of Time)*

relying on their own growing staff of engineers and scientists for information, Ramo and Wooldridge consulted outside experts from industry and academia on structural design, propulsion, guidance, reentry, and other subjects. Because Gardner was often preoccupied with other responsibilities and was not familiar with many technical details being discussed, Ramo frequently served as de facto chair of the Teapot Committee.[50]

In turning to a committee with heavy representation from leading science and engineering universities and from industrial corporations developing advanced technology, Gardner followed guidelines laid down years earlier by Hap Arnold, who said, "The next Air Force is going to be built around scientists—around mechanically minded fellows." Gardner also depended on the unprecedented prestige and influence at that time of leaders from the world of science and science-based engineering. Government, as well as the media and the public, turned to them for expertise on a wide range of government-policy and public issues. Responding to the prestige of science in the 1950s, many engineering schools added "science" or "applied science" to their names. In 1957, President Eisenhower recognized the status of science by naming James Killian, then president of MIT, the first presidential science adviser, special assistant to the president for science and technology.

From the military service's point of view, there was a downside to the influence of science advisers. They had the power to take policy making out of the hands of the military and shape strategy and projects in directions counter to current military preferences. We shall occasionally encounter this negative reaction as we follow the course of Atlas development, but the outstanding example remains the discrediting of Oppenheimer. Oppenheimer was the progenitor of the fission bomb and acknowledged spokesperson of the physics community on matters of national defense. His opposition to the development of the thermonuclear bomb placed a hurdle in the path of the Air Force when it moved to develop an intercontinental ballistic missile. Perhaps it was not entirely coincidental that in 1954 a government inquiry board denied Oppenheimer a continuation of his security clearance and disabled him as a science adviser.

In its initial report of February 1954, the committee focused on identifying alternative weapons systems capable of delivering a nuclear warhead over a distance of 5,000 to 6,000 miles. Besides ballistic missiles, the committee considered the possibility of an air-breathing cruise missile, such as Navaho or Snark. After deciding upon intercontinental ballistic missiles, the committee then concentrated on forecasting critical technical problems, or reverse salients, likely to arise in developing such a weapon.[51]

Most controversial of the committee's recommendations were those affecting Convair. While acknowledging the aircraft company's "pioneering work," the committee recommended radical modification in the Bossart team's missile design. The committee asked for one based "on a new and comprehensive weapons system study . . . adequately based on *fundamental science*" [my emphasis].[52] Informed that the weight of an atomic warhead could probably be reduced to as little as 1,500 pounds, the committee proposed reducing the total missile weight by about half, changing from the pressurized structure to a conventional self-supporting structure, and using a two-stage propulsion instead of a stage-and-a-half. Convair responded that its design still represented "the best that could be done," even with a lighter warhead.[53]

Gardner and the committee discovered that establishing an organization capable of presiding over the research and development of a system to produce the missile, an organization capable of transcending a cumbersome government bureaucracy threatening to hamper development in a "streamlined" manner, presented a challenge as great as, or even greater than, the technical ones.[54] Again the committee of scientists and high-technology engineers raised questions and created doubts about the effectiveness of Convair—and, in addition, they now raised doubts about the aircraft manufacturers in general.

The Teapot Committee recommended that the organization designated to manage the ICBM project take a systems approach— a recommendation strongly endorsed by Ramo, who had recently headed the electronics division and served as a vice president and director of operations at Hughes Aircraft. The committee believed that a systems approach would assure the earliest possible identification of the optimal design of a weapons system and achieve the earliest possible operational capability for an ICBM. Ramo later declared that "the Committee could hardly have called out a more definite lack of confidence in the specific Convair approach."[55]

As for the management organization, the committee recommended that a new organization—not Convair or another aircraft manufacturer—staffed by "unusually competent scientists and engineers" be set up to redirect the ICBM program and to exercise general technical and management control over all elements of the program, including research. The committee suggested that the government should be called on, if necessary, to "draft" the scientists, engineers, and managers needed to staff such an organization. These and other recommendations recall the World War II Office of Scientific Research and Development's mobilization of scientists and engineers.[56]

In raising substantial doubts about the ability of Convair and other large airframe manufacturers to manage an ICBM program, the committee and Gardner stirred up a hornet's nest. The entrenched prime-contractor approach relied upon an airframe manufacturer to design an airplane that would meet Air Force specifications, to supervise the design of subsystems by subcontractors, to take responsibility for keeping them on schedule, then to assemble the subsystems and flight-test the weapon. In the opinion of committee members, the manufacturers did not have the personnel and test facilities to fulfill such functions in designing and developing an ICBM; nor did they have a finely honed systems engineering approach comparable to the one taken, for instance, by AT&T or by Ramo's division at Hughes for building tightly knit electronic communications systems.[57]

The committee's concern about Convair's managerial competence resulted partially from the company's unwillingness to substantially modify its design for an intercontinental ballistic missile despite the acquisition of new data from background studies and scale-model tests, and despite the prediction that light nuclear warheads would be available and drastically alter design parameters, including accuracy requirements. The committee doubted, as well, that Convair, or any other airframe manufacturer, appreciated the role that electronics and computers should have in the designing and developing of an ICBM. In sum, the committee lacked confidence that Convair would be able to deploy a missile production system to assemble a sufficient number of operational missiles by the target date of 1962–63.

The committee discussed several alternatives to using an aircraft manufacturer as prime contractor. These included the formation of a consortium of several universities to establish a laboratory to preside over ICBM development. The formation in 1951 of MIT's Lincoln Laboratory to direct the development of the SAGE defense system offered a precedent for this course of action, as did the managerial and engineering functions provided earlier by the Jet Propulsion Laboratory. Another possibility involved recruiting civilian engineering and scientific talent through government civil service to form a technical direction and systems engineering staff for the project. Committee members also pondered the possibility of relying on a corporation like AT&T, with a strong management structure, to organize a division, or a subsidiary, to manage the project, as several industrial corporations had done for the Manhattan Project. Gardner, however, did not share committee members' enthusiasm for the performance of AT&T's Bell Laboratories as a systems engineer.[58]

Weighing the various alternatives, the committee felt that a university-sponsored laboratory might muster the academic scientists and engineers needed to design and develop missiles, but questioned such a laboratory's ability to preside over the vast industrial expansion needed for the production of missiles. (A few years later, as we have observed, MIT decided against having Lincoln Laboratory manage the production needed to deploy the SAGE air defense system.) Because government civil service did not offer the status that academics expected, and the remuneration that industrial engineers and managers expected, the committee doubted the success of a civil service approach.[59]

Having advised against these alternatives, committee members considered the capability of the newly formed Ramo-Wooldridge Corporation to act as systems engineer and to provide a large technical staff for the ICBM program; but, Ramo insisted—some persons later said, like Br'er Rabbit—that his newly formed company not be thrown into the systems engineering brier patch. Rejecting the Br'er Rabbit metaphor, he contends that he then preferred his company to design and build electronics hardware.

With the commitment to an intercontinental ballistic missile based on a redesign of the Convair Atlas and with the additional assurance—after the Operation Castle nuclear weapon tests in April 1954—that a light thermonuclear warhead could be designed for Atlas, the Teapot Committee in May 1954 reconstituted itself, renaming itself the ICBM Scientific Advisory Committee. Chaired by von Neumann, the group included seven from the original Teapot eleven plus two representatives from the nuclear weapons community—Bradbury of Los Alamos and York of Livermore. In addition, Charles Lindbergh, the famed transatlantic pilot, Jerrold Zacharias of MIT, and Frank Collbohm, president of the RAND Corporation, joined the committee. Earlier, we encountered Zacharias heading a summer study for the SAGE project and urging a holistic and systems approach to problem solving. Because of likely conflicts of interests, Ramo and Wooldridge withdrew as full members, but continued to serve as technical advisers.[60] With them off the committee, it was free to recommend major contractual responsibilities for their company.

Gardner aggressively prodded the Air Force Staff to take responsive and appropriate action. The possibility—even a veiled threat—that if the Air Force did not sponsor a crash ICBM program, the Eisenhower administration would create a Manhattan Project–type agency to get the job done greatly stimulated the Air Force to make a preemptive move.[61]

Air Force headquarters turned over the responsibility for the response to Gardner to the Air Research and Development Command, which had a strong pro-missile faction among its officers. Following reports from a subcommittee of von Kármán's Scientific Advisory Board and from a committee at the recently established Air Force Academy that the Air Force neglected research and development, the Air Research and Development Command had been established in 1950. Assigned responsibility for Air Force research and development, ARDC joined the older Air Material Command, which issued contracts for the procurement of weapons systems, to establish joint offices for each new major weapons system project involving a substantial research and development phase. In the past, the Air Material Command had greatly favored the development of bombers in contradistinction to long-range missiles.[62]

In the spring of 1954, Gardner proposed that the Air Research and Development Command take charge of an ICBM program to be headed by a general who would be closely advised by a military-civilian advisory group. Gardner recommended Major General James McCormack, vice commander of ARDC, and Brigadier General Bernard Schriever, assistant to the deputy chief of staff for development, who represented ARDC at Air Force headquarters. When McCormack retired because of illness, Schriever became the choice. He had favorably impressed Gardner when he served as an administrative aide to the Teapot Committee. In June 1954, the Air Research and Development Command established a Western Development Division (WDD), a field office on the West Coast that Schriever would command and which would have responsibility for managing the Atlas Project.[63]

]]]]]] Bernard Schriever: System Builder

A *New York Times* reporter described Schriever:

> . . . a 6 foot 2 . . . general who goes about his awesome task with the relaxed precision of a champion golfer sinking a one-foot putt . . . slim . . . a good looking man with alert but slightly bashful eyes, a straight nose and a chin that recedes without in the slightest suggesting weakness . . . a model of informality and gives the suggestion he has seen a lot of Jimmy Stewart films.[64]

Only forty-three years old when he took charge of the challenging, innovative project, Bernard "Benny" Schriever, born in Bremen,

Germany, had emigrated to the United States with his family during World War I. After his father died in an industrial accident, the family had difficulty making ends meet, but he was able to attend Texas A&M, where he earned an engineering degree and a reserve officer's commission in 1931.

During the Great Depression, he served as a reservist in the U.S. Army for a year, then won his Air Corps wings at Randolph Field, Texas. Schriever recalls that he was not in the "H.P. [Hot Pilot] category" so he found a home in a bombardment group "specializing in slow thinkers."[65] Between 1932 and 1938, he flew as a commercial pilot and acted as a camp director for the New Deal's Civilian Conservation Corps, where he learned a lot about command. In 1938, he received an regular Army commission.

Embarked on a regular Army career, Schriever married Dora Brett, daughter of Brigadier General George H. Brett of the Army Air Corps, and the couple moved to Wright Field, the center for Air Corps engineering research. Now focused on engineering, Schriever took a year of graduate study at the Wright Field Air Corps Engineering School and then another year of graduate study in mechanical engineering at Stanford University, earning a master's degree in 1942. He then served as chief of maintenance to the Nineteenth Bomber Group in Australia, where he also chose to fly thirty-eight combat missions. Later, as chief of maintenance for the Fifth Air Force, he was on hand during the liberation of Manila and had a six-week tour managing airfield development during the invasion of Okinawa. By war's end, at age thirty-three, he had risen to the rank of colonel.

Influenced by General Arnold's argument that the future of the Air Force depends on its effective utilization of science and engineering, Schriever chose a postwar assignment at the Pentagon. On one occasion, Arnold called Colonel Schriever into his office, urging him "to establish a very close relationship with the scientific community." Arnold reminded Schriever that World War I had been won by brawn, World War II through superior logistics, but that any future war would be won by brains.[66] A young officer in a service dominated by men who had had bomber and fighter com-

GENERAL BERNARD SCHRIEVER (CENTER), WITH MAJOR GENERAL ALAN BOYD (RIGHT) AND LIEUTENANT GENERAL WILLIAM IRVINE. *(Courtesy of Bernard Schriever)*

mands during the war, Schriever depended on the Air Force's future need for officers able to make research and development policy and manage R&D projects. At war's end, the likelihood that engineering officers would rise to the highest ranks seemed remote.

Assigned to Army Air Force headquarters, Schriever had primary responsibility for liaison with academic engineers and scientists, especially those who had helped design and develop such weapons as the proximity fuse, radar, the atomic bomb, and rockets during World War II. In this role, Schriever, an articulate and approachable man of keen intelligence, established close relations with such engineering and science leaders as von Kármán, von Neumann, and Teller. While he was attending the National War College, which prepares officers for broad policy-making positions, Schriever became known for complaining about "too much polishing of doorknobs instead of putting new technology to work." Returning to the Pentagon as assistant for development planning

BERNARD SCHRIEVER AT HIS DESK (NOTE CIVILIAN ATTIRE). *(Courtesy of Bernard Schriever)*

for the Air Force Staff, he became known as an ardent advocate of increased research and development. He introduced a systems approach to long-range planning that involved the analysis of potential military threats to the United States and a design for Air Force responses using advanced technology.[67]

By 1954, persons familiar with Schriever characterized him as having an excellent grounding in science and engineering, intimate knowledge of the workings of the Pentagon and the government as a whole, and an uncanny sense for evaluation of and "managing" people. After he took on the ICBM program management role, others found him intellectually acute, one perceptive scientist considering him an "intellectual giant."[68] President Eisenhower had to be with him only a few short times to be persuaded of his competence, objectivity, and dedication to the task.

When he assumed command of the Western Development Division in June 1954, Schriever inherited the von Neumann advisory committee and the problems on its agenda—including the challenge of designing a management structure that would set a precedent for military research and development projects. Schriever already had a potentially explosive problem on his hands, given committee members' strong feelings "that the Convair Company was not the best in the airframe business" to act as prime contractor for the Atlas Project. Believing that dropping Convair immediately as the design contractor would set the project back substantially, Schriever, advised by Ramo, delayed making a decision about the company's ultimate role.[69]

Schriever's initial inclination was to supplement and monitor Convair's engineers by having Ramo-Wooldridge assemble for his command a technical advisory staff of a hundred or so civilian engineers and scientists especially well informed about electronics and computers. They would provide Schriever with engineering background studies, advice on general project planning and direction, and evaluate the performance not only of Convair, but also of subcontractors. Thereby they would "keep industry participating in the program honest."[70]

The von Neumann committee strongly criticized Schriever's provisional plans at its July 1954 meeting. Committee members questioned whether or not Schriever, under his projected management structure, could retain sufficient authority. The committee urged Schriever to use the Ramo-Wooldridge organization as systems engineer and as a technical staff for the Atlas Project. Such a role would involve preparing an overarching conceptual design for the missile system, drawing up contractual specifications, coordinating and monitoring contractor performance, and performing evaluation testing. Airframe and other industrial companies would become associate contractors responsible for the development of various missile subsystems. The "other industrial companies" they had in mind were ones familiar with science-based technology, especially electronics and computers.[71] This scheme eliminated the traditional role of prime contractor—which Convair had hoped to play.

Committee members argued that Convair and other airframe companies could not provide the working environment, salaries, status, and organizational culture that would attract the best-trained and -qualified scientists and engineers. Ramo argued that airplane manufacturers could not attract outstanding professionals as long as the manufacturers continued to handle their engineers like factory workers by utilizing time clocks, whistles, rigid lunch hours, and the like.[72] The inflexible practices of engineering unions in large aerospace firms also concerned committee members.

Committee members assumed that the organization, headed by two Cal Tech graduates with doctoral degrees, could, in the position of systems engineer for the Atlas Project, attract engineers and scientists from Cal Tech, MIT, Bell Laboratories, and similar institutions that stressed the need for professionals trained in the physical sciences. Committee members believed that no single company, airframe or other, with the possible exception of Bell Laboratories, could muster the diverse intellectual resources to act as prime contractor.[73]

Since the early years of the aircraft industry, the airframe companies had acted as prime contractors, coordinating and monitoring the work of engine, electronics, and other subcontractors. Among the aerospace engineers, the aerodynamicists prevailed in the design process.[74] By recommending that Ramo-Wooldridge act as systems engineer, the committee was displacing engineers familiar with traditional aeronautical engineering practice with scientists and engineers grounded in the sciences and especially experienced in electronics and computers. To the committee, the airframe was merely a platform to carry complex, electronic guidance and fire control systems.[75]

The committee was calling for a revolution within the aerospace industry by drawing on American industrial and academic strength in electrical and electronic engineering to counter the strength of the Soviet Union in heavy engineering. Advantages that the Soviet Union might have in developing massive long-range missiles could be countered by the United States's dedication of its resources to developing more accurate and reliable missiles incorporating advanced electronics, especially guidance and control devices.

Following the July meeting, Gardner and committee member Lindbergh urged Schriever to accept the committee's recommendations to assign Ramo-Wooldridge full responsibility for systems engineering, thereby eliminating the need for a prime contractor, such as Convair. The committee even cautioned Schriever about using Convair as final-assembly contractor in view of its "previous position in the program."[76] He approved of using Ramo-Wooldridge with its "unique mechanism" for attracting leading scientists and engineers to supplement his small technical staff.[77]

The committee's recommendations caught Schriever and his skeleton staff of four Air Force officers in a bind.[78] High-ranking Air Force officers with whom Schriever consulted saw no need either for Ramo-Wooldridge as systems engineer to recruit scientists and engineers from outside the airplane-manufacturing industry or for abandoning the prime contractor approach.[79] Hearing this advice, Schriever had to expect not only disruptive, foot-dragging opposition from within the Air Force if he followed committee recommendations, but also loud complaints and disruptive difficulties from Convair and other aircraft companies who would be needed as subcontractors.

On the other hand, if Schriever failed to follow committee recommendations, he would antagonize civilian ICBM advocates in the Department of Defense, the university research establishment represented by von Neumann committee members, and nonaerospace companies at the cutting edge of technology who would also be needed as subcontractors. Thwarted, these people and organizations might argue "very convincingly at the highest levels" for a Manhattan Project–type of organization from outside the military to preside over the project.[80]

Schriever and his staff in August 1954 engaged in a formal study to answer the basic question: "Who will be responsible, repeat responsible, for the level of engineering decision usually exercised by a prime weapons system contractor?"[81] Schriever flew to Washington to consult Gardner, as well as Hyland, Lindbergh, Kistiakowsky, and von Neumann, committee members whom he considered the least "radical" and most enlightened about organizational matters.[82]

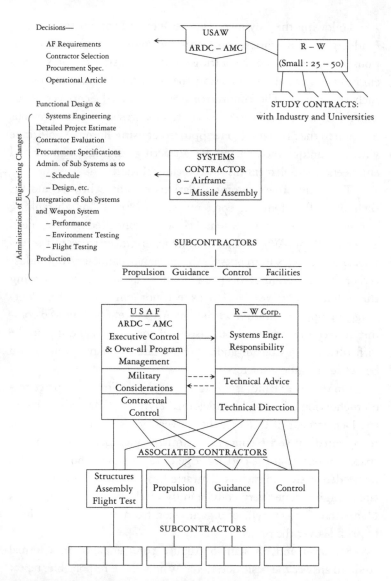

Decisions—

    AF Requirements
    Contractor Selection
    Procurement Spec.
    Operational Article

Functional Design &
    Systems Engineering
Detailed Project Estimate
Contractor Evaluation
Procurement Specifications
Admin. of Sub Systems as to
    – Schedule
    – Design, etc.
Integration of Sub Systems
and Weapon System
    – Performance
    – Environment Testing
    – Flight Testing
Production

Administration of Engineering Changes

USAW
ARDC – AMC

R – W
(Small : 25 – 50)

STUDY CONTRACTS:
with Industry and Universities

SYSTEMS
CONTRACTOR
o – Airframe
o – Missile Assembly

SUBCONTRACTORS

Propulsion   Guidance   Control   Facilities

U S A F
ARDC – AMC
Executive Control
& Over-all Program
Management
Military
Considerations
Contractual
Control

R – W Corp.

Systems Engr.
Responsibility

Technical Advice

Technical Direction

ASSOCIATED CONTRACTORS

Structures
Assembly
Flight Test

Propulsion

Guidance

Control

SUBCONTRACTORS

COMPARISON OF THE PRIME CONTRACTOR PLAN OF THE CONSOLIDATED VULTEE AIRCRAFT CORPORATION (CONVAIR) (TOP) AND THE SYSTEMS ENGINEER PLAN OF THE TEAPOT COMMITTEE AND BERNARD SCHRIEVER FOR ORGANIZATION OF THE WESTERN DEVELOPMENT DIVISION, 1954 (BOTTOM).
(From "A Study of the Development Management Organization for the Atlas Program, 18 August 1954."
Courtesy of Archives Ballistic Missile Organization)

These and other conversations persuaded him that the project had to have an innovative organizational structure able to coordinate unprecedentedly diverse engineering skills, scientific expertise, and industrial know-how in order to solve problems more complex even than those encountered during the Manhattan Project. Schriever also sought advice from both General Leslie Groves, who had headed the Manhattan Project, and J. Robert Oppenheimer.[83]

Schriever and his staff concluded that "the predominant technical aspects of this project have to do with systems engineering and with the close relationship of recent physics to all of engineering." Their report declared that a single industrial organization "generally lack[s] the across-the-board competence in the physical sciences to do the complex systems engineering job which the ICBM requires."[84] The advice of Gardner, Quarles, and the von Neumann committee prevailed. Systems engineers and systems engineering organizations would provide "technical generalship" because of "both the breadth of judgment and the level of decision" that they would exercise on the Atlas Project.[85] "Ranking military personnel" were not dominating decision making in a military-industrial-university complex.[86]

The Schriever report further declared that Ramo-Wooldridge, whose chief executives had "proven their competence in technical management of complex systems," would become a member of the Air Force "family," and would act as systems engineer for the project. Because Ramo had already persuaded "several full professors and university department heads" to take leaves of absence to work for the company, Schriever rested assured that the academic talent could be attracted.[87] He and his staff also decided that, in order to attract associated contractors, they should draw on a broad industrial base, one extending beyond the aircraft builders. Contracts would go to companies in the electronics and instrumentation industry for missile guidance and control, to companies in the propulsion industry for rocket engines, and to aircraft manufacturers for structure and final assembly. The organizational die was cast.

Schriever and his staff ended their report on a self-congratulatory note, concluding that their "organizational concept . . . will serve to accomplish the goals of the Atlas Project better than any other which can be devised" because "over-all management control" resides with the Air Force, because the use of associated contractors provides "the broadest industrial base," and because the scientists with "the nation's best brains" will be attracted to the project.[88]

]]]]]] Schriever's Managerial Style

Empowered to choose his staff from a pool of highly qualified Air Force officers, Schriever initially selected about a dozen officers with engineering and science backgrounds as well as administrative experience. This combination of qualifications provided them with the authority to monitor and question decisions made by the civilian engineers and scientists. Schriever chose Colonel Charles H. Terhune as his deputy commander of technical operations. Terhune's advanced engineering degree from Cal Tech, where he had been sent by the Air Corps as a young officer, and his three years as an engineering officer at the Air Corps's Wright Field, where he had charge of twenty-five officers designing aircraft in conjunction with aircraft companies, prepared him for the technical oversight and judgment Schriever needed from a deputy. Terhune's experience as an aide to Keller, whom President Truman in 1950 had asked to bring some order and discipline to the nation's dispersed guided missile program, also recommended him to Schriever. Additionally, Terhune's extensive flight experience in the Pacific during World War II gave him status among Air Force officers.

Schriever named other project officers to work in tandem with Ramo-Wooldridge project managers, who had responsibility for the development of various Atlas subsystems. The configuration of the missile determined the management structure: the guidance and the propulsion subsystems, for instance, each had its project officer and manager. The officers chosen by Schriever had engineering and science backgrounds comparable to Terhune's.[89]

Within months of having taken charge of the project, Schriever realized that he had to manage "outside the system."[90] By this, he meant the extricating of his command from the multilayered Air Force chain-of-command in order to gain direct access to the top funding offices and decision makers in the Department of Defense. Achieving this goal was analogous to designing a nose cone that could survive the complex return into the atmosphere. Schriever found that Air Force and Pentagon bureaucracy could overwhelm and delay his project with endless complications introduced at "many tiered approval levels of bureaucratic compliance for each special interest in the 'system' " and by a maze of budgetary review requirements. The persons who could say "no" but not "yes" blocked Schriever's managerial initiatives.[91] Delays in the construction of test facilities soon galled him for this threatened the opening phases of development. Under Air Force procedures, before funds could be appropriated a construction project had to be identified eighteen months ahead, in order to permit numerous reviews. In the Pentagon, Schriever lamented, swarmed a morass of assistant secretaries, "an R&D guy, money guy, installation guy, procurement guy . . . I would have to go to every one of those guys to test at Cape."[92]

Schriever prepared a "spaghetti chart" that graphically disclosed the tangled lines of communication and review that he and his staff had to toe in order to obtain the budget and acquire the resources needed to advance the ICBM program. Forty major offices and agencies with independent review responsibilities—in particular assistant secretaries in the Defense Department and in the office of the secretary of the air force—were the strands of a dense communication and review web.

Schriever laid out his chart before Gardner, saying that he "could not possibly get the . . . job done if I have to go through all this crap." Earlier, when urged by Gardner to take command of the WDD, Schriever accepted on condition that he could run the show "without any goddamn nitpicking from those sons-of-bitches at the Pentagon." He described his organization as "vertical" or project-oriented, while the Pentagon functioned "horizontally." Gardner empathized with a project director, focused on solving engineering problems, having to deal with administrators with economics, political science, and legal backgrounds who, in turn, were focused on the requirements of law, normative regulations, and budgetary constraints. Schriever later spoke despairingly of the influence of the "combination of lawyers and bureaucrats" and of a "system" constraining individual initiative.[93] He was lamenting the classic conflict epitomized by the confrontation of the engineer driven by technical and economic efficiency with the lawyer sensitive to interest-laden politics.

If Schriever had been directing the development of a low-priority program, he could not have "managed outside the system" no matter how great his powers of persuasion. Increasing intelligence information about missile activity in the Soviet Union, especially from radar installations in Turkey, however, helped give him the leverage, in the form of priorities, that he needed to push aside some routine. In March 1954, the Air Force secretary authorized "the maximum effort possible with no limitation as to funding" for the long-range missile program. Then in June the Atlas Project received the Air Force's top priority: the project would be "accelerated to the maximum extent that technological development will permit." These priorities, however, did not allow Schriever to manage outside the Air Force system. Furthermore, he could not avoid budgetary conflicts with other Air Force programs or sidestep "obstructive" Air Force review processes (and obstructive Air Force officers directing them).[94]

Because Schriever had "good support on the Hill," he and Gardner decided to use congressional leverage.[95] Engaging and persuasive in his appearances before congressional committees, Schriever had a sympathetic hearing from Senator Clinton P. Anderson, Democrat from New Mexico and chairman of the Joint Committee on Atomic Energy, and Senator Henry Jackson, Democrat from Washington State and chairman of that committee's Subcommittee on Military Applications, both of whom were alarmed by intelligence reports indicating that the Soviet Union might even be forging ahead in ICBM development. As a result of their inquiries, the senators sent a letter to President Eisenhower urging him to give the ICBM program overriding priority in the defense establishment, not just in the Air Force. They wrote that "although we fully understand the necessity for the detailed checks and rechecks which characterized Air Force procurement regulations, we believe that an overly rigid application of normal procurement policies to the ICBM is now slowing progress."[96]

Schriever also briefed the National Security Council, which subsequently recommended that the ICBM be designated a research program of the highest national priority, second to no others. Finally, in September 1955, President Eisenhower, taking into account the breakthrough in thermonuclear weapons and intelligence reports from the Soviet Union, acted accordingly. Armed with this priority, Gardner in September 1955 established an Air Force committee chaired by Hyde Gillette, civilian Air Force deputy for budget and program management, with the charge to evaluate the external management and control procedures governing the ICBM program. Schriever and several of his staff officers sat on the committee with other Air Force officers and with civilians. The committee's science advisory subcommittee included Ramo.[97] So loaded was the committee with WDD people that we should not be surprised that the recommendations of the Gillette committee reflected Schriever's wish to draw up long-term plans and budgets, make only annual presentations of these to the Pentagon and Congress, and deal only with a limited number of Pentagon people who could say "yes" as well as "no."

The Gillette committee submitted its proposals in October to the secretary of the Air Force. Reacting positively, he sent them on to the office of the secretary of defense which, following the committee's proposals, devised a management structure for the ICBM program that insured "that delays inherent in program review and administration are minimized."[98] The secretary of defense established a Defense Ballistic Missiles Committee and the Air Force established a corresponding Air Force Ballistic Missiles Committee. Schriever reported annually and directly to these encompassing budgeting and reviewing committees. Schriever's spaghetti chart became a historic document. Furthermore, both committees took advice from the Air Force's Scientific Advisory Committee, made up of engineers and scientists, not lawyers.[99]

Schriever not only managed to minimize the influence of Air Force and Pentagon middle management, he also adroitly side-stepped the politicization of decision making. Ramo provides a telling anecdote epitomizing Schriever's style. On one occasion, Secretary of the Air Force Harold Talbott importuned Gardner, Schriever, and Ramo to reconsider their decision to offer a rocket-engine contract to a West Coast company. Citing a Pentagon industry-dispersal policy and the concentration of defense contractors in the Los Angeles area, Talbott wanted the contract to go to a midwestern contractor who had lost out in the original competition, a contractor Ramo considered of marginal competence. Ramo believed that the contractor had used political influence to obtain Talbott's support and that dispersal policy was not the secretary's major motivation.

The secretary's proposal dismayed Gardner, Schriever, and Ramo. Yet each had compelling reasons not to challenge Talbott's decision: Gardner felt that, as a member of Talbott's staff, a negative signal from the secretary could quickly bring about his removal; Ramo's company depended on contracts from Talbott's Air Force; and Schriever "was an outstanding young general whose career might be ruined if he chose not to cooperate with the Secretary of the Air Force." Schriever had "the most to lose."[100]

Despite his vulnerability, Schriever rejoined that he could not comply with the request because his prior and overriding orders directed him to run his program so as to obtain an operational ICBM weapons system in the shortest possible time. Angrily, Talbott yelled at Schriever, "Before this meeting is over, General, there's going to be one more colonel in the Air Force!" Having given Schriever an order, he expected it to be obeyed. Schriever replied quietly that "perhaps Talbott might wish to put in writing an order specifically naming his choice of contractor . . . and at the same time . . . put industry dispersal above the need for speed in developing the ICBM."[101] Thinking more calmly over the impact in both industry and government of his issuing such an order, Talbott dropped his request. Ramo adds that not long afterward, Secretary of Defense Wilson requested Talbott's resignation because of his importuning government contractors to employ a firm he owned.

Schriever could not have faced down Talbott so coolly and successfully if the ICBM program had not been generally acknowledged to be a requisite for national survival in the Cold War missile race. With the end of the Cold War, managers of large-scale projects have become more attentive when congressmen expressed concern for jobs in their home districts.

Schriever, Ramo, and other Western Development Division managers and engineers risked making decisions without the protection of a red-tape bureaucracy that might later shelter them from unfriendly congressional investigations of mistakes and failures. A red-tape bureaucracy provides monumental amounts of paper defenses to surround critical decisions. Persons in responsible positions who may fear subsequent investigations secure from their superiors and subordinates documentation intended to demonstrate that circumstances dictated their policies and expenditures. Faceless circumstances, not they, are to be held accountable. General Leslie Groves, a risk taker who headed the Manhattan Project, joked that he would buy a house on Capitol Hill in Washington, D.C., after the war so he could walk to the investigative hearings of congressional committees.

]]]]]] Concurrent and Parallel Development

Cold War imperatives freed Schriever from bureaucratic red tape but also brought great pressure to complete the project at the earliest possible date. In response, Schriever introduced a policy of "concurrent development," or concurrency, which years later he characterized as a sine qua non for managing large weapons projects when "time is of the essence."[102] General Groves, during the Manhattan Project, and Admiral Hyman Rickover, in building a nuclear navy, had resorted to concurrent development earlier, and Schriever had discussed management problems with both of them. In the case of the Atlas Project, concurrence required simultaneous development not only of the missile with its various subsystems, but also of the facilities to manufacture and test the weapon, of the command and communication system for its operational control, and of a training system for the people who would operate the weapons system.

For instance, construction began on launching facilities at Cape Canaveral, Florida, before the engineers had a final design for the missile. As we shall soon learn, the concurrence approach needs highly developed management tools to insure coordination of concurrently proceeding and mutually dependent subprojects. Ramo recalls his concern, and that of others, that concurrent design and development could mean a massive concurrent failure. Yet, he and Schriever had no inclination to retreat to the slower and ostensibly safer series approach.[103]

The ICBM program also resorted to "parallel development," which further complicated the problem of coordination. The Manhattan Project had set a precedent for parallel development, as well, by developing two types of atom bombs simultaneously—the uranium and the plutonium. As early as August 1954, Schriever had ordered that there be a second source for missile engines, "for reasons of competition and the undesirability of entrusting such an important part of the program to one company."[104] The Rocketdyne division of North American Aviation had designed a stage-and-a-half propulsion system for the earlier Convair design that had been carried over into the redesigned Atlas. Following Schriever's call for parallel development, the Western Development Division engineers contracted with Aerojet-General Corporation to design a two-stage propulsion system.

Then in the spring of 1955, the Western Development Division chose parallel contractors not only for propulsion but for each of the major subsystems. It decided to assemble the second set of subsystems as the "Titan." The Glenn L. Martin Company had charge of final assembly of the Titan and therefore ran in parallel with Convair, which by then had the contract for final assembly of

BERNARD SCHRIEVER (CENTER) AND SIMON RAMO (RIGHT) AT THE DEDICATION OF THE WESTERN
DEVELOPMENT DIVISION BUILDING, 1955. *(Courtesy of Simon Ramo)*

the Atlas.[105] The Ramo-Wooldridge engineers serving as Schriever's
supplemental technical staff exerted even greater influence in
designing and developing the Titan than they did on the Atlas.[106]

Schriever and Ramo defended this costly duplication, which
assured that the failure of one contractor would not delay the project and insured a competitive search for solutions to front-edge
research and development problems.

As if the plate of the Western Development Division were not
full enough in 1955, the Department of Defense and the Air Force
in December ordered the command to preside over the design of an
intermediate-range ballistic missile (IRBM) as well as the two
intercontinental ballistic missiles. WDD selected Douglas Aircraft
Company as the airframe and assembly contractor for the 1,200-
mile-range Thor missile. The Navy would also develop an intermediate-range missile, the solid-fuel Polaris, which was launched
from submerged nuclear-powered submarines.[107]

	ATLAS	TITAN
Airframe	Convair	Martin
Guidance		
Radio-inertial	General Electric	Bell Telephone
All-inertial	A.C. Spark Plug	American Bosch and MIT
Propulsion	North American	Aerojet General
Nose Cone	General Electric	AVCO
Computer	Burroughs	Remington Rand

ATLAS AND TITAN CONTRACTORS. *(Courtesy of Davis Dyer and TRW Corporation)*

]]]]]] Simon Ramo

Like Schriever, Ramo emerges from the Atlas history as a pivotal figure. Although neither man was a heroic system builder who handed down decisions from hierarchical peaks, each acted within an infinitely complex web of circumstances, some inherited from the past, others emerging with the developing program. While Schriever represents a new type of Air Force officer who feels confident of his ability to manage complex technology projects, one who moves at ease with eminent scientists and engineers, Ramo epitomizes the scientist-engineer who moves into the aerospace industry with its ingrained managerial style, then introduces a high-technology, science-based, systems-oriented managerial approach.[108]

Ramo had a California Institute of Technology doctorate and California-itis—an ailment that "causes its victims to be in extreme pain unless they reside in Southern California." "Smog, the definitive cure," Ramo recalls, had not yet been discovered when he settled in California.[109] Born in 1913, he spent his first twenty years in Salt Lake City, where his father had emigrated as a boy from Brest Litovsk, then a part of Russian Poland. His mother, too, had come to the United States from Russia. In Salt Lake City, the Ramos flourished in a small Jewish community unburdened by prejudice from a predominantly Mormon society—the Mormons had suffered from prejudice themselves. Young Ramo found Mormon and Jewish cultivation of education and music stimulating. Doing well in high school, an accomplished violinist, he became concertmaster of the orchestra, an achievement that furthered his career in science and engineering.[110]

He chose to study engineering at the University of Utah, hoping that the courses would not conflict with his music. Since science, mathematics, and analysis of engineering problems came easy to him, he found time to play as a soloist with the Utah Symphony Orchestra. Looking back on engineering school, Ramo finds its focus on technical problem solving and its handbook approach narrow and stifling of creativity. Discovering that his humanities and social science courses at Utah required no study other than attending the lecture, he "hardly even cracked the book."[111] He faults the engineering faculty, then and now, for not encouraging students to see the relationship of technology and society and for not recognizing the social values embedded in technology.

Forgoing a career as a musician, he decided to obtain a doctorate in engineering. One of his professors sent his alma mater, Cal Tech, a strong letter of support and helped Ramo, already ranked first in a class of several hundred engineers, win a modest scholarship. He would have preferred a school like MIT, which he then considered more focused than Cal Tech on applied, as contrasted with pure, science. The Cal Tech emphasis on theoretical foundations soon persuaded him, however, that an engineer must be grounded in physics, and he was pleased to find his electrical engineering courses almost identical to those taken for the degree in physics. In an advanced electrical engineering course, he also learned that problems from the field of practice could not be solved if one observed disciplinary boundaries—"the kind of thing that is, even today," Ramo observes, "usually not taught in universities."[112] Commitment to theory and interdisciplinarity remained defining characteristics of Ramo the engineer.

Because of his virtuosity as a violinist, Ramo became the concertmaster of the Cal Tech orchestra. He accompanied Cal Tech President Millikan to teas for wealthy San Marino dowagers who might be persuaded to endow Cal Tech more handsomely if they believed that the institution trained the whole man. Upon completion of three years of doctoral studies in electrical engineering and physics at Cal Tech, Ramo hoped for a position at the General Electric Research Laboratory. Fortunately for him, the company interviewer valued the well-rounded student highly, or as Millikan said, "the whole man." Ramo's violin performance on an occasion arranged by Millikan for the GE recruiter, and Ramo's highly original approach to problems, stood him in good stead. General Electric offered him a position with the promise that he would enjoy Schenectady—the company base—because of its fine symphony orchestra.

Ramo may have been surprised as well as pleased because at Cal Tech he had been told that the large companies, such as General Electric, Westinghouse, and American Telephone and Telegraph (AT&T), did not hire Jews.[113] General Electric, he had also heard, hired engineering graduates from the midwestern universities, young men who molded more easily into a GE image, socializing more easily on teams than engineers from such elite schools as MIT and Cal Tech.

Although he found the GE-required training courses devoid of theory, within a few months the organizers of the courses, recognizing his thorough grounding in physics, asked him to revise the courses. This endeavor eventually culminated in his first book, *Fields and Waves in Modern Radio* (1944), a seminal text on electronics including microwaves and semiconductors that has remained in print, with sales to date of more than a million copies.[114] After a year or so, the company moved Ramo into the General Engineering Laboratory, where he investigated high-frequency microwaves, electron optics, and high voltages, fields quite distant from GE's traditional business of manufacturing generators, engines, and transformers.

After war began in 1939, Ramo's research on microwaves became classified as an essential problem area for the development of military radar. Because the MIT Radiation Laboratory microwave program soon overwhelmed his small-scale endeavor, Ramo wanted to take a wartime leave to join the MIT laboratory. The director of the GE laboratory persuaded him that it would be a mistake to join a laboratory run by professors that would be involved in theory and not in serious weapons work. The real microwave work, he confidently predicted, would be done in industry at places like General Electric. Within a year, Ramo found the reverse true. He decided that large "companies . . . become bureaucratic and arrogant, and miss new trends, leaving opportunities for new and small firms."[115] He dreamed of establishing a company in California modeled on David Packard's Hewlett-Packard Company and Arnold Beckman's Beckman Instruments.

As the war drew to a close, Stanford, the University of California–Los Angeles (UCLA), the University of California–Berkeley, and Cal Tech, aware of his publications and pioneer work in the microwave field, each offered him a position, as did the Douglas, North American, and Lockheed companies. Stricken by California-itis, Ramo considered the offers of these aircraft companies, but their emphasis on manufacturing and their dependence on the National Advisory Committee for Aeronautics for research curtailed his interest. On the other hand, an offer from the Hughes Aircraft Company attracted him.

Owned by the eccentric multimillionaire Howard Hughes, the aircraft company, located in Culver City, California, did not have a reputation for research and development and had few government contracts as compared with the large aircraft companies. Yet from Ramo's point of view, the relatively new and small company, unstructured and backed by enormous wealth, offered opportunities for an enterprising and highly imaginative person.

Leaving General Electric after almost a decade to take the Hughes position, Ramo quickly obtained a sizable Air Force cost-plus contract for the development of an air-to-air missile later named the Falcon. As head of the electronics division, he recruited a host of talented scientists and engineers from universities and industry, especially persons with California-itis. "All of a sudden," Ramo remembers, "there was a clear embryo of what became in a matter of two to three years the largest concentration of technical-degree people working in stuff of this kind in the country."[116] Only

Bell Laboratories, Ramo believes, had a larger pool of scientific and engineering talent.

Ramo persuaded those whom he wished to recruit that his division's concentration on electronics, computers, and the systems approach tied into a technological revolution that was expanding human brain power much as the previous technological revolution had increased muscle power.[117] He also argued that his company dedicated its considerable resources to advanced research and development while well-established enterprises such as General Electric and RCA focused upon less technically demanding and interesting commercial projects. Able to charge facilities against contractual budgets, the Ramo division could soon boast of excellently equipped laboratories as well. Ramo assured potential recruits to his division who expressed concern about the influence of the eccentric Hughes upon the division that "he made no decisions about it. He did not conceive it or participate in it or follow it."[118]

Wooldridge, Ramo's good friend from the Cal Tech student days, left Bell Labs and became Ramo's associate director of research. The two men thought so much alike that they could even complete each other's sentences. Wooldridge brought with him not only his wartime experience in designing control devices for weapons but his familiarity with Bell's systems approach to problem solving, a variation of which he and Ramo instituted in the electronics division.

The division soon installed complex electronic systems on interceptor aircraft, systems including air-to-air missiles, radar, computers, cockpit displays, and communications. Ramo intended only to fabricate prototypes to prove concepts and designs, leaving serial production to large companies. He found, however, that large manufacturing subcontractors set up production schedules too slowly, so Hughes began manufacturing the electronic systems as well as designing them. Ramo and his colleagues began to see the airplanes of the traditional manufacturers as simply platforms for their electronic systems. He confidently anticipated that electronics would become the tail that wagged the airplane dog. Robert E. Gross, chairman of Lockheed, tried to buy the Hughes electronics division, saying to Ramo: "We can't ignore the electronic side of things and just simply build frames around equipment that you guys serve up. Eventually we're going to have to be a complete systems group."[119]

By 1950, most of the Hughes company's 15,000 employees were working for the electronics division. Ramo had become company vice president for operations and, in effect, Hughes's general manager. "After about three or four years," Ramo recollects, "it was perfectly clear that ours was one of the big, remarkable postwar phenomena. There was no other company postwar with this kind of a buildup of business."[120]

Despite its growth, his division, Ramo maintains, did not become swollen, slow-moving, and bureaucratic. For this, he gives credit to his systems engineering approach. Air Force procurement contracts became projects presided over by interdisciplinary systems engineering groups. Project groups met frequently to identify the subsystem interface problems, then to negotiate compromise solutions among the subsystem engineers, each pushing to decrease the complexity, cost, and weight and to increase the reliability of the respective subsystem for which he or she was responsible. Ramo gives an example:

> The rocket engine guy says, "I can't swivel this cockeyed rocket to change direction. It's too hard to make a device that will swivel and do it fast and with big angles. So you've got to let me get by with just a tiny little bit of motion here." And the guidance expert says, "I can't allow that. By the time I have figured out what I need to have to do it's a little late for you to get it corrected with such a little bit of motion. You've got to give me a bigger motion." Well, those guys fight and they draw curves and they negotiate.[121]

Ramo acted as a lead systems engineer, mediating and stimulating but urging realistic, incremental compromises to allow a project to move ahead in the face of never-ending problems. His readiness to acknowledge that there is no one best solution to most problems heightened his effectiveness as a moderator and negotiator among persons with a variety of enthusiastically proposed solutions.

To motivate and reward the handpicked engineers and scientists, Ramo and Wooldridge paid above-average wages and provided working conditions better than those common in the aircraft industry. His division provided individual offices (ten-by-twelve cubicles)—with telephones. The management called weekly "town meetings" where scientists and engineers discussed the general problems and opportunities of the workplace and the company.

Despite the achievements of the electronics division, Ramo, Wooldridge, and a number of their leading engineers left the Hughes Company in 1953. Howard Hughes's increasingly eccentric, unpredictable—even unreliable—behavior in other ventures in which he had invested heavily, such as Trans World Airlines and movie production companies, caused concern in the Defense Department that the possibility existed of his intervening and adversely affecting his aircraft company, a supplier upon whom the government depended for critical products. Defense Department heads such as Wilson, Talbott, and Gardner repeatedly hinted to Ramo that they would look with favor on the establishing, by a few key people, of another company in the same field to compete with Hughes Aircraft.

Deciding to establish a new company, Ramo and Wooldridge chose as a sponsor Thompson Products, a Cleveland, Ohio, precision engineering and manufacturing company that supplied the aircraft and automotive industries. Thompson wanted to expand into electronics. From this alliance emerged the Ramo-Wooldridge Corporation in September 1953. Less gregarious and more strongly analytical than Ramo, Wooldridge became president, leaving Ramo free to direct projects that he found more interesting than administrative and financial matters.

The young company, with a secretary, an administrative assistant, and two telephones, established its Los Angeles–area headquarters in a recently vacated barbershop. A long line of engineers seeking jobs, from Hughes and elsewhere, overwhelmed the tiny organization on its opening day. A few days later, the Air Force gave the company a letter contract calling for a background study of the research and development needed to initiate an intercontinental ballistic missile program.

Not until January 1955, however, did the Air Force provide a formal contract assigning systems engineering and technical direction responsibilities to the guided missile research division of Ramo-Wooldridge.[122] The division and Ramo, however, had become fully involved in the ICBM program since the spring of 1954 on the assumption that a formal contract would be forthcoming. The 1955 cost-plus-fixed-fee contract provided Ramo-Wooldridge with a fee of 14 percent, higher than usual in the aircraft industry. The Air Force added contract clauses banning the Ramo-Wooldridge Company from using the guided missile research division's privileged access to information about the research and development activities of ICBM associate contractors. This, the Air Force argued, could be used by the company to produce competing hardware.

Providing systems engineering and technical direction for the project, Ramo and his engineers needed to invent an organizational structure that allowed close interaction between them and the officers on Schriever's staff. Thus they set up parallel organizational structures with a project manager from Ramo-Wooldridge paired with a project officer from Schriever's staff. Project officers and project managers together presided over the five project groups: guidance and control, aerodynamics and structures, propulsion, flight test and instrumentation, and weapons systems analysis.

The physical layout of the WDD headquarters building in Inglewood, California, mirrored the close interaction of Schriever's and Ramo's staffs. Schriever and Ramo, located in adjacent offices, met almost every day. Their deputies, Dunn of Ramo-Wooldridge, formerly of the Jet Propulsion Laboratory, and Terhune of the WDD, also had adjacent offices. Because the officers on Schriever's staff wore civilian clothes for security reasons, uniforms did not set up a communication barrier.

Ramo and Schriever functioned well in tandem because neither had a short fuse or gave in to emotional outbursts; neither tried to dominate the relationship or to take sole credit for achievements;

both considered management a matter of common sense—and the taking advantage of fortuitous circumstances; and both unflinchingly made decisions under pressure. Ramo admired Schriever for not using the ICBM program as a means to further his own career and for seeing the project as an end in itself, an end for which a career might be sacrificed if need be.[123]

Ramo speaks glowingly of the engineers and scientists he handpicked for the ballistic missiles project. He "marvels" at the accomplishments of his top forty or so engineers, who had an average of twenty years of field experience in science and engineering. To dramatize their achievement, he would display a board with his leading engineers' names in a great circle. He would then spin an arrow, listing the notable achievements of the person upon whose name the arrow came to rest, thus demonstrating the even spread of talent at his disposal.

The subsequent careers of his leading engineers support his claim. William Duke became executive vice president of ITT; Richard DeLauer became undersecretary of defense; James Fletcher became director of the National Aeronautics and Space Administration (NASA); Ruben Mettler became chief executive of the Ramo-Wooldridge successor firm, Thompson Ramo Wooldridge (TRW); George Mueller later directed the Apollo manned lunar-landing program; and Albert Wheelon became chairman of Hughes Aircraft. Once again, as in the case of the Whirlwind project, the development of a large-scale system acted as a fertile learning experience for a host of engineers.

]]]]]] Systems Engineering

To appreciate the importance of systems engineering, we should try to imagine what a formidable management problem one faces when confronted with the problem of scheduling and coordinating hundreds of contractors developing hundreds—even thousands—of subsystems that eventually must be meshed into a total system. When development depends on research, scheduling becomes enormously complicated: the development of many subsystems depends on the prior design, research, development, and testing of others. With research usually unpredictable, scheduling becomes a virtual nightmare. The preparing and serving of a five-course dinner provides an everyday analogy with the scheduling and coordinating of projects like Atlas. Some ingredients cannot be prepared until others are ready; the courses must be brought to the table in sequence and piping hot. Such preparation also entails the systematic planning and coordinating of the purchase of food; of house, linen, and silver cleaning; of design of table setting and flower arrangements; and of the congenial grouping of carefully chosen guests.

Ramo, the system builder, and his corporation claim a number of managerial innovations that stemmed from their involvement in the ICBM program, especially from their cultivation of the art and science of systems engineering. While Ramo argues that Ramo-Wooldridge decidedly shaped systems engineering as practiced in the field today, he does not contend that systems engineering began during the program or even during this century. He, as do many other engineers, points out that the system builders of the great pyramids practiced systems engineering as they scheduled and coordinated countless construction activities. In the twentieth century, as Ramo also acknowledges, designers and constructors of telephone and electric power systems practiced systems engineering, with several of the theoretically inclined rationalizing the process. As we have noted, Schriever promoted a systems approach to planning even before meeting Ramo.

Ramo and his engineers made an original contribution to the development of systems engineering by creating an organization dedicated to scheduling and coordinating the activities of a large number of contractors engaged in research and development and in testing the components and subsystems that are eventually assembled into a coherent system.[124] The builders of the pyramids at Giza did not, we assume, award research and development contracts nor were the utility companies that managed the development of telephone and electric power networks focused primarily on systems engineering as was Ramo's guided missile research division.

Asked why he focuses on systems engineering as the key to

managerial success, Ramo refers to his preference for the interdisciplinary approach to problem solving as contrasted to the specialized, disciplinary one taken by many other engineers. He enjoys looking for interconnections.

His self-characterization is reminiscent of philosopher Isaiah Berlin's commentary on a line from the Greek poet Archilochus: "The fox knows many things, but the hedgehog knows one big thing." "There exists," Berlin continues, "a great chasm between those, on one side, who relate everything to a single central vision, one system less or more coherent . . . and, on the other side, those who pursue many ends, often unrelated and even contradictory. . . ."[125] Ramo, however, undoubtedly prefers to be known as a systems engineer rather than a hedgehog.

Ramo contends that systems engineers see interconnections and preside over projects. They communicate a central vision to the specialized engineers. If governed by a central vision, Ramo believes, engineers are more likely to negotiate and to make specification trade-offs among subsystems. If unwilling to make trade-offs, engineers are liable to distort the overall system they are developing.

Having delineated the subsystems and components that constitute the overall system, the systems engineers prepare preliminary, or conceptual, designs and do background studies. They give special attention to defining preliminary specifications, or quantifiable characteristics such as weight, size, and energy input or output, especially specifications at the interface of subsystems and components. Systems engineers speak of "monitoring the interfaces." If these should be changed during the project, then the systems engineers change and adapt the specifications of all affected subsystems and components in order to maintain a harmonious coordination. We should recall that changes in one component of a tightly coupled system usually reverberate throughout the system in a domino effect.

Scheduling the development and research to be done by the contractors responsible for the myriad subsystems and components becomes the messiest chore for the systems engineer. Final specifications for some subsystems and components cannot be laid down

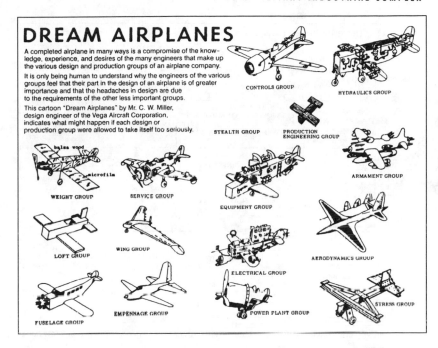

DREAM AIRPLANES

A completed airplane in many ways is a compromise of the knowledge, experience, and desires of the many engineers that make up the various design and production groups of an airplane company.

It is only being human to understand why the engineers of the various groups feel that their part in the design of an airplane is of greater importance and that the headaches in design are due to the requirements of the other less important groups.

This cartoon "Dream Airplanes" by Mr. C. W. Miller, design engineer of the Vega Aircraft Corporation, indicates what might happen if each design or production group were allowed to take itself too seriously.

CONTROLS GROUP

HYDRAULICS GROUP

STEALTH GROUP

PRODUCTION ENGINEERING GROUP

ARMAMENT GROUP

WEIGHT GROUP

SERVICE GROUP

EQUIPMENT GROUP

AERODYNAMICS GROUP

LOFT GROUP

WING GROUP

ELECTRICAL GROUP

FUSELAGE GROUP

EMPENNAGE GROUP

POWER PLANT GROUP

STRESS GROUP

WHEN A SYSTEMS ENGINEER DOES NOT NEGOTIATE DESIGN COMPROMISES, ENGINEERS DESIGN THEIR DREAM AIRPLANES. *(Drawing by C. W. Miller. Courtesy of Stanley Weiss)*

until the specifications for some other related subsystems and components have been decided upon, and these specifications often cannot be determined until extensive research and testing have been done.[126] These realities help explain why initial designs and their specifications are usually called preliminary. Systems engineers look forward to the time when they can change the classification of specifications from preliminary to semifixed by incorporating them in working drawings used in the manufacture of prototypes of missile components and subsystems.

Systems engineers seek to "optimize" overall system characteristics, such as the reliability, range, and accuracy of a missile. Optimization, a more complex concept than efficiency, results from trade-offs. Optimization can result from trading off, or reducing, the capacity of the steering component of a missile guidance system by designing a more precise guidance-control component. Similarly, the development of a lighter warhead makes possible a trade-

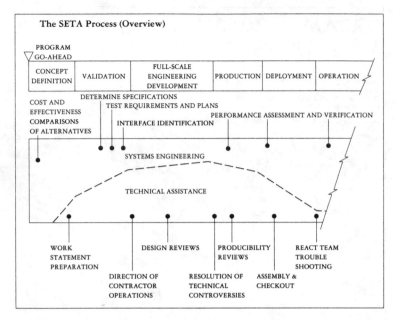

THE SYSTEMS ENGINEER/TECHNICAL ASSISTANCE PROCESS. *(Courtesy of Davis Dyer and TRW Corporation)*

off that reduces the propulsion, or thrust, requirement. Ramo took as his responsibility deciding which system specifications or characteristics should be optimized, then persuading the subsystem and component engineers to negotiate with one another to this end.

In the ICBM program, once the specifications were laid down in working drawings, the systems engineers performed not only as systems engineers but also as technical directors. In this role, they cooperated closely with Schriever's staff of officers in deciding upon contractors and in closely monitoring their performances. They took part in the testing of both missile subsystems and components as well as the assembled missiles. Then they advised the Air Force whether or not individual missiles should be purchased.

In performing the critical task of choosing ICBM contractors, Schriever reminded his officers and engineers who might have become impatient with the long-drawn-out process that "literally billions of dollars of business would be involved."[127] Ultimately, WDD rank-ordered the potential contractors, then invited the top firms to bid. Before this took place, however, a painstaking selection process of almost three months ensued.

The selection of a nose cone contractor is a good example of this process. Initially, a group of officers and Air Force civilians, long familiar with Air Force contractors' previous performances, gathered at Wright Field to discuss the qualifications of various companies for developing a nose cone. Lieutenant Colonel Otto Glasser, a member of Schriever's staff who organized the meeting, asked those present not to discuss aircraft manufacturers because "we are already quite familiar with their capabilities and qualifications."[128] Earlier, Convair had been found "not uniquely qualified to provide all components for completion of the nose cone . . . and pretty damn swamped with the rest of the job—not top talent."[129] Preliminary studies also led to the conclusion that no single contractor had the necessary competence in thermodynamics and aerodynamics, so the contractor chosen would need to draw on the "really high powered people" in the universities and industrial research laboratories.[130]

The group made two lists, one for technical skills and the other for management capabilities, ranking the companies as excellent, good, fair, and X (unqualified). After evaluation of technical skills, Oliver Aircraft, Ford Motor Company, Allis-Chalmers, and Minnesota Mining were eliminated from the competition. General Electric, Bendix Aviation, and NAA Aerophysics, a division of North American Aviation, emerged as leading contenders. Bendix, General Electric, and NAA Aerophysics ranked high when judged against the criterion of management capabilities. Paradoxically, in view of its subsequent selection as a nose cone contractor, AVCO Manufacturing Corporation received low ratings in metallurgy and systems management, being judged "not too attractive a contractor" for nose cone development.[131]

After lengthy discussions with engineers and WDD officers, Glasser made his final recommendations. He found Borg-Warner and AVCO weak in systems engineering. General Motors did not rank as a major contender because "of treating their subsidiaries on a competitive basis and of looking always at the production dollar."

Chrysler won praise for its ingenuity and "engineering advancement" but lacked the technical competence needed for nose cone development. Glasser decided that a contract to NAA Aerophysics would incur "tremendous political repercussions from the airframe contractors" because of North American's other Air Force contracts. Glasser concluded that only General Electric, Westinghouse, and Bendix Aviation should be considered further, and he ranked them in that order.[132]

He believed that one of the aircraft companies, which, we recall, had not been evaluated by the group at Wright Field, should have the contract to develop an interim nose cone while General Electric, Westinghouse, or Bendix Aviation should be given a sole-source contract to design a greatly improved nose cone. After consultation with other Air Force commands, Schriever and Ramo, with their staffs, chose General Electric as nose cone contractor for the Atlas missile and AVCO for the Titan missile.

]]]]]] Black Saturdays

To contend with the daunting task of coordinating the contractor's activities, Schriever and Ramo instituted "Black Saturday" discussions. These involved Ramo, Schriever, Ramo's project managers, Schriever's project officers, and others invited in accord with the problems on the agenda. Attention usually focused on failures to meet scheduled development milestones or, as Schriever quipped, "inch stones," and on the failure of components to meet performance standards during tests. Cost overruns also demanded considerable attention.

Before the high-level Black Saturday meetings, Ramo and his staff at Ramo-Wooldridge usually held a preliminary overview meeting on the preceding Wednesday and, even before this gathering, they often met with engineers from associate contractors. Ramo could then on Saturdays focus on technical problems, while Schriever and his staff could direct their primary attention to costs.[133] "I didn't try to get into every damn little detail," Schriever recalls, "we had plenty of problems . . . we busted up a lot of missiles . . . we had a lot of failures, but we were going full bent."[134] He did not want to hear of successes, only of problems.

When Black Saturdays revealed seemingly intractable problems, Ramo often brought in consulting experts from industry and academia. Members of the von Neumann scientific advisory committee proved to be particularly rich sources of assistance. When, for instance, the oxygen supply to the Atlas engines was shutting off too soon, a call for help went out to an ad hoc committee of five experts. Charles Lauritsen of Cal Tech, a member of the original von Neumann committee, scrutinized drawings of the missile and located a point where he deduced that excessive pressure had built up. Recommending a precisely located hole to relieve the pressure, he solved the problem. The missile reentry problem also required frequent consultation with outside experts.

Black Saturdays took place in a specially designed project control room located in a massive reinforced concrete vault at WDD headquarters. Staffed by an Air Force officer and several assistants, the control room displayed countless charts showing progress—or nonprogress—toward "inch stones," tests results, itemized costs; red flags singled out critical problems that had to be solved.[135] Technical and expenditure information flowed into the control room from thousands of contractors, but checking the accuracy of this information remained a problem. In order to stay off the critical problems list, contractors tended to present their progress optimistically.

By 1957, Schriever's officers and Ramo's engineers were overseeing more than 200 contractors who, in turn, had thousands of subcontractors. The number of engineers and scientists assigned to the ICBM project rose from about 50 in 1954 to about 800 a year later. They supplemented the small staff of about 150 officers and civilians on Schriever's staff. By 1957, the numbers had risen respectively to about 2,000 and 700.

By the mid-1950s, an early digital computer in the control room displayed real-time information from the countless contractors. Schriever believes that the ICBM may have been the first military program to use digital computers to process control room information. Colonel Terhune, Schriever's deputy for technical matters who had been involved in a number of aircraft development projects, believes that ICBM may also have been the first military program to depend on a control room.[136]

]]]]]] Nose Cone Development

Of all the engineering problems encountered during the Atlas project, the design of the nose cone, or reentry vehicle, proved most baffling. The von Neumann committee had insisted that the ICBM project required more high technology and science input than a single aircraft company possessed. Solving the reentry problem consumed as much as 11 percent of the Atlas development budget.[137] One of Schriever's officers wrote in 1956 that "in the entire IRBM/ICBM program, the re-entry nose cone represents the one major R&D problem. . . . there does not exist today any proven method of re-entering the earth's atmosphere, particularly at the speeds involved and insuring safe delivery to a point in space or to a point on the earth's surface."[138]

The nose cone, the section of the missile carrying the nuclear warhead, separates from the body of the missile a minute or so after liftoff. It returns to the atmosphere, following a ballistic flight of about 5,000 miles, at a velocity of more than 10,000 miles per hour. The interaction between the reentering nose cone and the atmosphere generates destructive conditions not encountered previously in the study of gas dynamics. Earlier, airplane designers, in their search for structural materials that would withstand high temperatures, assumed that air could be treated as a compressible fluid. But at supersonic velocity, air behaves more like a radiant incompressible fluid. As the temperature of the air approaches that at the surface of the sun, molecules tend to break down. Engineers, realizing that aluminum would not survive these conditions, considered using exotic materials such as titanium for the body of the nose cone. Furthermore, they sought methods to dissipate, or transfer away, heat at the surface of the nose cone, thereby shielding the body of the nose cone.

Designers sought various ways to transfer surface heat, including using a heat sink (relying on the physical principle of conduction as solid-surface material vaporizes) and depending on ablation (employing layers of material that absorb energy as friction causes them to burn away). The engineers had to determine what material to use in a heat sink located at the front of the nose cone and what material to use on the surface of the nose cone to dissipate heat by ablation. Not only did choice of material affect heat transfer, but the aerodynamical shape of the heat sink and the nose cone did as well.

Background studies led engineers and scientists, in conjunction with their academic consultants, to conclude that sufficient theoretical knowledge of copper heat-sink behavior existed to permit design of a nose cone for full-scale test missiles. On the other hand, they concluded, theoretical knowledge of ablation was insufficient to support the design of a nose cone dependent on ablation. In the case of ablation, experiment—or testing—had to precede design. In other words, experience had to come before theory. Engineering is not simply applied science; know-how often precedes know-why.

Testing occurs at many stages during technological development.[139] In general, testing moves from mathematical and computer simulation of an artifact, or system, functioning in its environment through testing of small-scale physical models to full-scale testing of an assembled subsystem or system. After the system turns operational, testing continues so that postinnovational modifications can be made.

Often, testing provides a reality check for theoretical hypotheses. This not infrequently results in revised theory as well as design changes. Ramo's engineers and Schriever's officers took part in designing and monitoring the testing procedures carried out by General Electric and AVCO, the two nose cone contractors. Believing that competition stimulated innovation, the Western Development Division had each of the contractors proceed independently in designing and testing their respective nose cones. They exchanged a minimum of information.

A well-developed network of existing government, university, and industrial test installations facilitated missile testing, including the Jet Propulsion Laboratory's twenty-inch supersonic tunnel and the Naval Ordnance Laboratory's hypersonic wind tunnel. Newly installed mainframe computers at the various test sites aided in analysis of data and in testing by simulations.

Because existing wind tunnels had limited capacity for simulating the conditions of nose cone reentry, however, the Atlas team stimulated—and often funded—the installation of other experimental techniques and facilities, including shock tubes at AVCO's research laboratories and the X-17 reentry test vehicle at the missile systems division of the Lockheed Corporation. The shock tube makes measurements from stationary models of missile components, such as the nose cone, by simulating extremely high temperature conditions for intervals of a few thousandths of a second. These tests helped the engineers predict the changing state of the atmospheric environment that would surround the nose cone during reentry. One test engineer found shock tube results pertaining to boundary layer transition, ionization, and recombination so impressive that he anticipated designing a nose cone without "burdensome allowances for ignorance"—allowances more commonly called fudge factors.[140] Excess weight in the copper sink constituted a particularly burdensome fudge.

The first stage of the three-stage X-17 reentry test vehicle drove the 10,000-pound device out of the atmosphere. After the vehicle reversed direction, the second-stage and third-stage engines drove it back into the atmosphere at a velocity approaching that of a missile nose cone. Designing the X-17 telemetering devices for relaying information to test engineers about ambient and material conditions also proved challenging. (We should note that engineers allowed nature to perform one test—they examined the condition of meteorites that had survived similar experiences.)

Ramo relied upon a panel of consultants from academia and government research institutes to advise periodically on the nose cone problem. In 1956, having concluded that his team needed a "fresh look," he convened a panel headed by Robert Bacher of Cal

Tech to look for fresh approaches to the reentry problem. The panel included persons not expert in the practical problems of nose cone design but generally informed about gas dynamics and materials.[141]

The panel provided "basic information." WDD engineers and scientists defined this as a "gathering of analytical or experimental knowledge which is applicable to nose cones, but may be more general in nature—perhaps applicable to high velocity aerodynamics in general." In addition, the Bacher panel contributed to "advanced design," defined as pertaining "to a particular type of advanced nose cone."[142]

Elsewhere in the project's network of researchers and designers, about three-quarters of the eighty or so General Electric engineers working on the heat-sink project as of mid-1957 concentrated on the acquisition of basic information. The remaining quarter did advanced design. General Electric expected to spend $6.3 million in the following eighteen months on the nose cone project. In the case of AVCO, about thirty engineers and scientists made up its nose cone group at its Lawrence, Massachusetts, facility and about sixty more did nose cone work at its Everett, Massachusetts, laboratory. Most of them acquired basic information through testing and analysis. The nose cone team depended heavily on computer studies to generate information that would allow advising and monitoring the contractors.

General Electric's management of the nose cone project did not please either Schriever's officers or Ramo's engineers. One officer faulted General Electric for not implementing a "project manager" style of management and for not creating a "streamlined organization," or division, dedicated solely to the nose cone project—as had been agreed. The WDD team wanted the GE project manager who headed the "streamlined organization" to have call on the "best people" from within and without the company and to resort to every facility at the company's disposal. Unless GE introduced project management, one of his staff officers informed Schriever, nose cone "problems will not be solved in the time span of this program."[143] The officer also criticized GE for not following suggestions given by Simon Ramo.

Even after GE named a project manager, Schriever remained concerned about the situation at the company. So he, with several of his officers and engineers, met with company executives. At the meeting, Dunn of Ramo-Wooldridge hinted that the GE project manager might not have the competence and drive to do the job. Others argued that GE depended too much on subcontracting hardware development, that some GE-designed and -manufactured hardware would not meet specifications, that cost control was not tight enough, that GE had been slow to acquire and utilize "top-notch" technical personnel in certain critical areas, and that GE test schedules had not been realistic.

On the basis of its computer studies, engineers found the initial design of the GE engineers for a nose cone "completely technically unsound." After Ramo-Wooldridge "directed" GE to do a new design using computer-generated values for critical parameters, the company arrived at what engineers judged to be a rational design for a nose cone, which, incidentally, had more than one thousand components. Because of delays in the scheduled development of test missiles, the technically unsound nose cone could be abandoned and design of a better one initiated without concern that this would delay full-scale testing. These criticisms and directives suggest why contractors and subcontractors sometimes complained about interference from the "so-called experts" who monitored their activities.

As a result of continuing meetings, the GE performance improved, in the opinion of Schriever and Ramo's team. Ramo-Wooldridge did continue to perform "technical audits of the more critical components."[144] Schriever even considered ordering a "management survey" of General Electric, but the continuation of improved performance changed his mind.

On the positive side, Dunn reported in the spring of 1957 that the establishment of a missile and ordnance systems department and of a nose cone program manager at GE's Philadelphia plant resulted in more "aggressive and intelligently inventive" nose cone work. He also commended the company for an expeditious release of drawings to manufacture in January 1957 and for the significant results being obtained from testing as the project progressed toward scheduled nose cone flight testing in October.[145]

By June 1958, seventeen heat-sink nose cones had been tested with wind tunnels, the X-17, and test missile flights. These tests verified nose heat transfer, dynamic stability, pressure distribution, drag calculations, and other parameters affecting design of both the heat-sink and the ablation nose cones. Confident about the test data, engineers recommended moving from design and production of heat-sink nose cones to design and production of the theoretically superior ablation for subsequent use on both the Atlas and Titan missiles.[146]

]]]]]] Downgrading Convair

Before observing the flight testing of ICBM missiles and their operational deployment, we should consider the reaction of the aircraft industry, especially the Convair company, to Ramo-Wooldridge acting as systems engineer, thereby displacing the customary prime contractor, and thus breaking an aircraft-industry practice honed during World War II. Convair, we should remember, had expected to be named prime contractor for the Atlas Project.

The Air Force decision favoring Ramo-Wooldridge caused resentment not only at Convair but throughout the industry. Schriever had received an early warning of the negative reaction when Frank Collbohm took exception in 1954 to the committee's recommendations that Ramo-Wooldridge become the systems engineer. Before becoming president of RAND and a member of the von Neumann committee, Collbohm had been an engineer and executive at Douglas Aircraft, a major airplane manufacturer.

Collbohm argued impressively, logically, and complexly. He pointed out that in 1953, RAND studies had shown that an increase in bomb yields and a consequent reduced need for missile accuracy would make possible an operational ICBM by 1960, earlier than previously predicted. He based this sanguine forecast on the assumption that the Air Force, depending on industrially experienced aeronautical engineers, would not try to develop the most advanced performance missile possible, but would trade off performance for the early operation capability. He expressed concern that, on the other hand, inexperienced engineers and scientists recruited by Ramo would prove overly committed to advanced technology and perfection of design. Their design for a missile might result in more advanced performance but would delay the development of a satisfactory operational missile.

Collbohm contended that a modified Convair design for the Atlas should be turned over immediately to industrial contractors for engineering and production. Although he wanted an airframe contractor to have the prime contractor and attendant systems responsibility, he judged Convair not to be the best airframe contractor for the program.[147] "In my opinion," he added, "the only way in which we will mobilize the best talents and resources from industry for the job is to allow full play to the forces of industry competition. . . . It seems unthinkable at this late date to by-pass these qualified existing organizations and to start from scratch to build up, as the technical staff of the Project Office, a new organization for this purpose."[148]

Behind the complaints of the uneasy aircraft industry, we find not only adherence to the prime-contractor concept but concern that the new management plans at WDD would give Schriever and Ramo the kind of micromanagement control that Admiral Rickover had successfully introduced during the nuclear navy project. Schriever countered that the airframe companies did not have the in-house capacity to design and manufacture the major missile subsystems, including propulsion, electronics, and guidance and control. Furthermore, he predicted that the leading nonaircraft companies with expertise in these fields would be unhappy as subcontractors to aircraft companies whose technology and management style seemed antiquated.[149] The aircraft companies, we should add, saw the nonaircraft companies staffed by Young Turks who lacked the tacit knowledge required to perform effectively in a long-established and complex industry.

Schriever also responded to a rash of industry communications in trade journals "extolling the virtues" of Convair while denigrating WDD and Ramo-Wooldridge.[150] Industry spokesmen attacked the Air Force for subsidizing Ramo-Wooldridge's hiring of large numbers of engineers and scientists at wages above the industry average and building up its physical facilities at the same time. With comparable support, the manufacturers argued, they could greatly increase their own capability in electronics and the other fields needed for missile development. Schriever countered that this was an example of the "pot calling the kettle black" in that "the airframe industry owes its existence and present affluence to Government support in contracts and Government facilities." He cited as examples the $100 million of government-owned facilities used by Douglas, the $80 million used by North American, and the $22 million used by Lockheed. He also noted the unimpressive record of the aircraft manufacturers in developing the Navaho, Snark, and Matador missiles.[151]

"The Air Force is convinced," Schriever maintained, "that the technical complexities and the advances of the ICBM are each substantially greater than past development projects."[152] "What the hell capability," he now asks, "did they [Convair] have to be a prime contractor. . . . could they marry [integrate] the warhead?"[153] The systems engineering approach, he asserts, really developed a whole new industry in this country, which even the disgruntled contractors now recognize as viable. He gravely doubts in retrospect that the project would have developed well if "we had given Convair a prime contract."[154]

With the WDD management structure in place, Convair had to ship off its nose cone drawings that it expected to transform into hardware to AVCO and to turn over responsibility for missile guidance to GE. Ramo and Schriever judged these tasks to be especially demanding of scientific and high-technology competence.[155]

Having expected to be the prime contractor and to perform the systems engineering function, Convair found itself reduced initially to the design and development of the missile structure. Convair had reason to take exception to a decision that left it, a company with a long record as a prime contractor, with little more responsibility than developing and manufacturing missile structures to which some engineers pejoratively referred as "tanks" or as housing for propulsion and electronics.

Convair managers pointed out that a Convair team had created the highly innovative breakthrough ICBM design in the late 1940s, and that the company had lobbied the Pentagon in the early 1950s to sustain interest in an ICBM. A lack of Air Force enthusiasm for long-range missiles as well as lack of funds for prototypes and testing, not Convair's resistance to change, explains the slow evolution of the Atlas design, Convair spokespersons argued. Convair characterized those members of the von Neumann committee who considered Convair weak in electronics and systems engineering as "ivory tower experts with no feel for the realities of the technical problems involved."[156]

When engineers, many of whom had never designed an aircraft, appeared at Convair headquarters in San Diego in 1954 bearing new specifications for the Atlas missile, Convair, after looking them over, decided that the Ramo-Wooldridge engineers, many with backgrounds in electronics, had little understanding of structural dynamics and strength of materials. Sharp disagreements about design specifications arose because of the differing backgrounds, perspectives, and priorities of the Convair and Ramo-Wooldridge engineers.[157] Some of the less opinionated engineers at Convair found the experts very smart and well intentioned but woefully lacking in tacit knowledge and experience.[158]

By contract, Ramo-Wooldridge engineers could require Convair to introduce design changes by having Schriever issue technical directions (TDs) to Convair. Yet Ramo-Wooldridge rarely resorted to this weapon, because issuing a TD allowed the contractor to claim a contractual change and charge for the increased costs incurred. In time, as understanding and tolerance increased between the systems engineers and contractor engineers, they negotiated differences, thereby avoiding contractual design revisions.

Relations improved in late 1954 when Schriever contracted with Convair not only for the missile structure but also for the assembling and testing of the missile. In addition, Convair received a contract to develop and manufacture the Azusa guidance system designed for the earlier Atlas. Numerous test missiles eventually used Azusa. In general, Convair's influence on the design and development of Atlas increased as Ramo-Wooldridge focused its systems engineering and technical direction energies on Atlas's twin, the Titan missile.

In general, Convair responded to criticism from those who doubted its broad engineering capability by increasing its resources in those areas in which the von Neumann committee had found it lacking. Retired Air Force General Joseph McNarney, head of Convair, assured Schriever that he would substantially increase the number of engineers assigned to Atlas and that many of these, some from academia and government laboratories, would have electronics backgrounds. The company also greatly enhanced its computer design capabilities.

Convair, in addition, improved the working conditions and the salary level of its engineers. It set up a scientific advisory committee for its Atlas Project, intending to use consultants and committees much as Ramo-Wooldridge had done. Ramo's disdainful remarks about the demeaning circumstances under which engineers had to work at the airplane companies seemed no longer to apply to Convair. After the company built a new building for its engineers, some of them believed their work environment to be better than the conditions under which their counterparts at Ramo-Wooldridge worked. Convair had responded to a wake-up call.[159]

]]]]]] Testing, Failures, and Deployment

Before concluding our account of the Atlas Project, we shall briefly summarize the testing of full-scale missiles and the deployment of operational Atlas and Titan missiles. In passing, we should note that our account stressing research, design, and development does not deal with the manufacture of the missiles by Convair and other contractors.

Convair and Ramo-Wooldridge engineers along with monitoring and budget-watching Air Force officers took charge of the flight testing beginning in 1956 at Cape Canaveral of the first series of ten full-scale missiles, the A series. These were built only for testing. Weights substituted for the nose cones; several other operational components were simply left off. The tests generated an enormous amount of data from the telemetering of the missiles in flight. Fifty radio channels transmitted information signals to the ground from sensing instruments placed at 150 points on each test missile. From the analysis of the Atlas A tests, the engineers introduced design changes in Atlas B, which had all of the features of operational missiles. Analysis of test data from the flights of these missiles led to modifications in the Atlas C missiles, some of which were operationally deployed. The Air Force moved to full deployment with the Atlas E missiles.[160]

The testing procedure involved initial ground testing of components to see that they fulfilled interface specifications. Many of the components, especially those laden with electronics, came in black metal boxes. These had output signals that had to conform to the interface specifications laid down by the systems engineers during the design phase. So it was not necessary to open the black box—a procedure that led to the use of "black box" as a metaphor, meaning that only the output of a component, not its inner working, is of interest to the engineers assembling a system. After ground testing, the assembled missile was flight tested. Test results often led to modifications of the missile design, but such changes had to be cleared with project headquarters because changes to a component caused ripples throughout the system.[161]

Before full deployment, the engineers experienced many test failures as well as encouraging successes. By late 1958, an Atlas nose cone had flown 6,000 miles and landed accurately. (The Russians may have had a similar success about six months earlier.) To counter charges that the United States had fallen behind the Soviet Union and that there was a missile gap, President Eisenhower reported to the Congress in January 1960 that by then there had been fifteen launches with ranges of more than 5,000 miles and accuracy within 2 miles, a performance superior to the Russian.[162] Yet about half of seventy-seven U.S. test launchings between June 1957 and February 1961 failed to meet expectations.[163]

Because the media, the public, and some congressmen did not understand that testing of design changes often involved failures, the Air Force and the Western Development Division experienced sharp criticism. Schriever reacted philosophically. "When at the leading edge of technology and ploughing new ground . . . if you do not have failure every now and then, you are not taking enough risks." He recalls that after five or six uninterrupted successful launchings, von Kármán, an experienced and innovative engineer, sent him a telegram: BENNY. YOU MUST NOT BE TAKING ENOUGH RISKS. "Today," Schriever observes, "if you have a failure, you are fired. People will not take risks."[164]

By 1961, Convair was assembling about three Atlas D missiles each month.[165] Some months later the Glenn L. Martin Company began to assemble the Titan and the Air Force to deploy it. Following a maximum deployment of ninety-nine missiles on station in 1964, the Atlas was gradually phased out, transformed into a space launch vehicle with a long career.[166]

]]]]]] Windup

After Atlas, Schriever continued as head of the Air Force's long-range-missile program, presiding over the development of the solid-fuel Minuteman that displaced the Atlas and Titan missiles. After becoming head of the Air Force Ballistic Missiles Division and then the Air Force Systems Command, successor organizations to the Western Development Division, he became head of the Air Research and Development Command. After retiring as a four-star general, he continues to advise industry and government on complex management problems. We shall discuss one of his major projects in the following chapter.

Because aerospace companies contended that the systems engineering division of Ramo-Wooldridge could pass on proprietary information from aerospace contractors, with which the division worked, to the profit-making manufacturing divisions of the corporation, the Air Force in 1960 turned over its missile projects systems engineering responsibilities to the newly formed Aerospace Corporation, a nonprofit corporation.[167] Ramo became executive vice president and chairman of the executive committee of TRW, Inc., which absorbed the Ramo-Wooldridge Corporation. Ramo serves as adviser to government and private organizations about science and technology policy in addition to writing often about the subject. The engineering and management community has bestowed many awards upon him.

As both Ramo and Schriever look back on the Atlas years, they undoubtedly express the nostalgia and pride of achievement of most of the officers, engineers, and scientists who worked beside them. Both Ramo and Schriever frame their memories of the ICBM program in heroic mold: they and their associates took high risks; they ventured into the unknown; their enthusiasm carried them through discouraging failures; and they had authority commensurate with their responsibilities.

Ramo, acknowledging that the performances of the twenty or so principal industrial contractors, the hundreds of second-tier contractors, and the thousands of third-tier contractors was not perfect, contends that the ICBM program remained "remarkably free of major cost overruns, schedule slippages, waste, and fraud" as compared to comparable programs decades later.[168]

Schriever takes understandable satisfaction in pointing out the enormous complexity with which his command had to cope: the total engineering development effort for the B-52 bomber required about 7 million engineering man-hours; the Atlas/Titan Project required about thirty-one million. Yet development of the B-52 took eight years, while the Atlas/Titan Project took four and a quarter years from initial hardware contracting in January 1955 to design freeze in March 1959.[169]

The circumstances conditioning large-scale weapons projects today differ greatly, both Ramo and Schriever believe, from those that they experienced: "It is 'business as usual' today in the government; it was 'mission impossible' then on the ICBM program in the decade from 1954 to 1964." Ramo believes that with the passing of the Cold War commitment the government will not be able again to set up "the tight, closely integrated systems management that proved so effective on the ICBM program."[170]

Yet the systems engineering approach as practiced during the Atlas Project has spread to other industry and government projects. We shall next consider the spread of the systems approach in the 1960s.

IV Spread of the Systems Approach

Engineering systems are designed from the inside outward, that is, from components into a functioning whole. Behavior of a system is a consequence of interaction of its parts, parts that themselves must be understood and interconnected.

Jay Forrester, *System Dynamics*

▼ In the 1960s, the systems approach, or systems science, spread from military-funded SAGE and Atlas into the civil areas of government and into industry.[1] Not only SAGE and Atlas engineers and scientists cultivated systems science as an approach to technical and scientific problem solving, but managers who took part in these projects were eager to apply systems science to solving managerial problems as well.[2]

The 1960s witnessed not only the spread of the systems approach but the flood tide in the postwar world of the influence of science and high-technology experts in government and industry.[3] Systems enthusiasts, especially management experts and social scientists, successfully promoted the introduction of the systems approach into the administrative hierarchies of the U.S. Department of Defense, under Robert McNamara, and the civil government, especially in connection with the Great Society programs of the administration of President Lyndon B. Johnson.

Americans are reluctant to acknowledge the countless ways in which military experiences, both in peacetime and wartime, shape their national character and behavior. Only a few historians have chronicled the transfer of technology and management techniques from the military to the nonmilitary sectors. Both the early-nineteenth-century "American system of production" and early-twentieth-century "scientific management" took root in the military-funded sector.[4] The spread of the systems approach presents an even more persuasive example of the military origins of major American institutions and activities. We should qualify this generalization by reminding ourselves that the military-industrial-university complex, not the military alone, created large-scale projects like SAGE and Atlas.[5]

The advocates and enthusiasts of the systems approach often did not agree. Some had systems engineering in mind; others were thinking of operations research; still others focused on systems analysis. Usually, those speaking of systems engineering had in mind either the management of the design and development of systems with purely technical components, such as the Whirlwind computer, or sociotechnical systems with both technical and organizational components, such as the Atlas project.[6] Proponents of operations research referred usually to quantitative techniques used to analyze deployed military and industrial systems. Those expert in systems analysis compared, contrasted, and evaluated proposed projects, especially those that would create weapons systems. Frequently, the definitions overlapped and, in practice, usage varied from these definitional norms. In general, advocates of the systems approach perceived, conceived of, or created a world made up of systems.

Countless technological enthusiasts, however, used systems terminology with only the haziest idea of its meanings. They hoped to gain authority and add weight to their recommendations and decisions by associating them with the systems approach.[7] This approach carried considerable cachet because of the demonstrable successes of engineers and managers who used systems engineering, operations research, and systems analysis to analyze and construct weapons systems.

The systems approach spread rapidly during this period because gullible enthusiasts as well as experts believed that it would make possible the creation and control of the vast and complex human-built technical and sociotechnical systems that structure the industrialized world.[8] "Man must learn to deal with complexity, organized and disorganized, in some rational way," argued one advocate who expressed the commitment among engineers, scientists, and managers who had tried and found the systems approach effective in responding to physical and organizational complexity, especially in connection with military-funded projects.[9] So emboldened, the experts and their disciples confidently, even rashly, tackled a multitude of the complex technical and social problems plaguing American society in the 1960s.

Among the complex problems, none seemed more pressing than the social and physical dilemmas of the large cities. In his reference to the space program that carried on the systems engineering approach introduced during the Atlas Project, Vice President Hubert Humphrey eloquently stated the widespread faith in the systems approach when he said in 1968:

> . . . the techniques that are going to put a man on the Moon are going to be exactly the techniques that we are going to need to clean up our cities: the management techniques that are involved, the coordination of government and business, of scientist and engineer. . . . the systems analysis that we have used in our space and aeronautics program—this is the approach that the modern city of America is going to need if it's going to become a livable social institution. So maybe we've been pioneering in space only to save ourselves on Earth. As a matter of fact, maybe the nation that puts a man on the Moon is the nation that will put man on his feet first right here on Earth.[10]

]]]]]] Systems Engineering

During the 1950s, tens of thousands of America's most highly trained, experienced, and dedicated engineers, scientists, and managers garnered memorable learning experiences from creating SAGE and constructing Atlas. Generous salaries from prestigious companies and agencies attracted outstanding graduates from the leading schools of engineering and science, as well as experienced professionals from other activities, into defense work. Persuaded that the technological threat from the Soviet Union was genuine, and committed to the proposition that the defensive and offensive weapons they were designing, developing, and deploying were a response to a real and present danger, they worked long, hard, and ingeniously to solve managerial, technical, and scientific problems. Solving problems lastingly influenced their professional styles, their ways of defining and approaching problems. As much or more prestige would accrue to an engineer's being successful on a major weapons system project as would his having graduated from a leading engineering school.

A nation does not walk away unchanged from its large technological projects. The investment of resources is one thing; the lasting influence of these projects is another. Traces of the SAGE experience, for instance, can be found in the way computer engineers and software designers approach and solve problems today. The billions of dollars expended on SAGE and Atlas provided the engineers, scientists, and managers physical and personnel resources of quantity and quality unprecedented in the long history of technology. Such support emboldened them to tackle enormously complicated problems. The creative problem solving they did for SAGE and Atlas proved an exhilarating experience for many of them. The urgency and secrecy surrounding the projects heightened the excitement and pride of achievement.

Not only individuals became carriers of the SAGE and Atlas learning experiences, but organizations did as well. The RAND Corporation, the Lincoln Laboratory, the MITRE Corporation, the Ramo-Wooldridge Corporation, the System Development Corporation, the Air Corps Systems Command, and the Navy Special Projects Office mushroomed from the soil of the large-project 1950s. The unique confluence of military, political, economic, and technological circumstances in the 1950s brought a decade of managerial and technological achievement that has lastingly influenced the character of American engineering and management.

Computer-based control of large and complex processes emerged from SAGE; refined means of scheduling and coordinating the countless subsystems and components that make up massive projects and programs came out of Atlas. SAGE and Atlas utilized and reinforced the transdisciplinary approach. Teams of engineers, technicians, and scientists polarized around problems rather than disciplines. As a result, new discipline-transcending organizational forms, such as Lincoln Laboratory, Ramo-Wooldridge, and the Special Projects Office presided over system-building projects rather than discipline-bound departments. The transdisciplinary team approach is still considered front-edge management almost a half century later. Together these innovations constitute a milestone in the history of management—the management of things and also the management of engineers and scientists.

By 1964, at least nine institutions, including MIT, Carnegie Institute of Technology, the Johns Hopkins University, and the University of Pennsylvania, were offering systems engineering programs that combined systems engineering and operations research courses. Subject matter in courses and publications included optimization, analog models, probability and statistics, information and communication theory, game theory, queueing theory, control and servo theory, linear programming, and computer programming.[11] Many professors of systems engineering confidently asserted that their subject matter constituted a new academic discipline; others called it a management science.

The use of the appellation "systems engineer" became increasingly common. By 1970, there were an estimated 11,000 systems engineers, most of whom worked in the aerospace industry.[12] Alexander Kossiakoff, a scientist and engineer at the Applied Physics Laboratory of Johns Hopkins University who designed weapons projects, defines the systems engineer as the prime mover among engineers working on large systems projects.[13] Seeing an analogy between a military campaign and the management of a large systems project, he characterizes the systems engineer as a technical general who exercises a breadth of judgment and a high level of decision making that extends from the conception of a system through its development and deployment.[14] The systems engineer understands the language of specialists and coordinates the team.

]]]]]] Operations Research

Operations research, or operational research, as the British refer to the managerial technique, originated in England before World War II. A small number of distinguished British scientists, most of whom leaned to the political left, take credit for having created operational research. It is variously defined:

> ... the application of scientific methods, techniques and tools to the problems involving the operation of a system so as to provide those in control of the system with optimum solutions to the problems.[15]

Or:

> ... as the study of administrative systems pursued in the same scientific manner in which systems in physics, chemistry and biology are studied in the natural sciences.[16]

Or:

> ... as a scientific method of providing executive departments with a quantitative basis for decisions regarding the operations under their control.[17]

These definitions of operations research (OR) rest on the assumption that there is a readily and easily definable scientific method or scientific manner. Since this is questionable, we shall define operations—or operational—research by historical cases, or examples.

Solly Zuckerman, a British scientist who early in his career focused on anatomy, states in his biography that operational research took shape during dinner meetings that he organized in the late 1930s for a group of congenial, prominent, politically left-leaning scientists.[18] Their number included P. M. S. Blackett, a physicist who would later win the Nobel Prize; J. D. Bernal, X-ray crystallographer, spokesman for the scientific left, and author of

Science in History; J. B. S. Haldane, geneticist and mathematician; Joseph Needham, X-ray crystallographer and historian of Chinese science; Robert Watson-Watt, radar pioneer; and C. H. Waddington, embryologist and geneticist. They believed that science had momentous social impact and that scientists should take responsibility for its character.

In 1937, Watson-Watt organized a program to systematically coordinate an air defense system that included as components a coastal radar network, interceptor aircraft, and antiaircraft artillery. The Royal Air Force called the team "the operational research section." After Britain and the Soviet Union became allies in World War II, others of the Zuckerman group dedicated themselves to the development of operational research in offensive operations as well as defensive ones against German air and submarine attack. Bernal saw this as the initiation of a successful effort to join physical science, engineering, and practice into one common, articulated discipline useful in war and peacetime industry.[19] Transdisciplinarity becomes a hallmark of operational and operations research.

In 1941, Blackett organized for the Army an operational research group to coordinate antiaircraft defenses and for the Coastal Command a group to optimize defenses against submarines and mines. Stressing a transdisciplinary approach, he organized another OR group that included three physiologists, two mathematical physicists, one astrophysicist, one Army officer, one surveyor, one general physicist, and two mathematicians. The group became known as "Blackett's circus."[20] In 1942, Blackett became the director of naval operational research for the Admiralty. In order to systematize the use of depth charges against submarines, the Navy OR team analyzed the history of antisubmarine operations, focusing on such variables as the number, the explosive capacity, the depth of detonation, and the damage inflicted by depth charges during attack.[21] As in this project, operational research practitioners often compiled and analyzed historical case studies. In essence, OR specialists gathered data about past operations in order to plan future ones; they functioned as applied historians, relying when possible on quantitative data and statistical analysis. History became a laboratory of experience.

The Royal Air Force employed operational research first to analyze the history of British air operations over Germany and then to suggest tactics on the basis of this analysis. Teams of British physicists, mathematicians, social scientists, and historians analyzed the interaction of such variables as the altitude, speed, number, and formation of bombers as well as the characteristics of the German defenses, British and German airplane losses, and target damage. Weather conditions and the amount of light or darkness also entered into the equations. On the basis of the analysis, the OR teams recommended tactics to the bomber commands.

Cecil Gordon, a social biologist and another of the OR pioneers, drawing on his prewar work with drosophilas, used an analogy between the life cycle of the fruit fly and the optimal maintenance cycle of aircraft patrolling the Bay of Biscay in search of submarines. In this way, he managed to stagger aircraft maintenance, giving preference in the queue to those airplanes whose damage would take the least time to repair, thereby insuring that some airplanes were always conducting aerial reconnaissance despite the limited number of airplanes available.

The socially concerned scientists who promoted OR for defense before and during the war contended that OR had an important role in revitalizing British industrial production and government services afterward. The rise to power of the Labour Party gave them ample opportunity to test their assumptions. Herbert Morrison, who had occasionally joined the scientists at Zuckerman's dinners, became lord president of the council in the Labour government and head of the Department of Scientific and Industrial Research. From these influential positions, he advocated the use of scientists, including those familiar with OR, in the designing of industrial and social policy.[22]

Gordon drew up reports on OR for a government Committee on Industrial Productivity (CIP); Bernal brought the OR approach to the housing section of the Ministry of Works; and Blackett, his influence enhanced by the award of a Nobel Prize in 1948, organized a series of conferences on operational research attended by hundreds of industrialists. The advocates of OR also introduced the study of the subject into the universities.

Only modest results followed on these enthusiastic endeavors, however. Civil servants turned over policy making to outsiders from science reluctantly at best, no matter how distinguished and well meaning the scientists. Industrialists hesitated to invest in well-salaried OR teams, especially when OR seemed to many of them little more than a variation of scientific management practices that they had long tried.

]]]]]] Operations Research in the United States

During the war, operations research gained a foothold in the United States through the activities of a British-American network of scientists and military officers. The British encouraged American physicists and naval officers to introduce OR. The transplanting of OR from Britain to the United States recalls the comparable transfer of the magnetron tube, which greatly enhanced the effectiveness of American radar, and the transfer of the Maud Committee Report, which persuaded the Americans of the feasibility of developing an atomic bomb.[23]

Philip M. Morse, an MIT research physicist who headed the Brookhaven National Laboratory and the Defense Department's Weapons Systems Evaluation Group after the war, had close contact with his friend Blackett during the war. Morse persuaded high-ranking U.S. naval officers that OR would be effective in planning convoy tactics for defense against, and destruction of, German U-boats. Impressed by the highly regarded MIT professor, the Navy asked him in to form and chair an Antisubmarine Warfare Operations Research Group (ASWORG) that would report to higher echelons and thus avoid having its advice lost in the chain of command. Morse also obtained the authority to obtain information about operations in the field rather than having to depend on second-hand information. He gathered about him other civilian scientists and specialists, including physicists, mathematicians, theoretical geneticists, and insurance actuaries.

Before turning to the history of postwar operations research, however, we should note the development of OR in the U.S. Army Air Force during the war. The Air Force named the OR function "operations analysis." Made aware of the Royal Air Force's effective use of operational research, the American Joint New Weapons Committee headed by Vannevar Bush, the MIT electrical engineer who mobilized American science and engineering during the war, asked W. Barton Leach, a Harvard law professor, and Ward F. Davidson, an industrial scientist, to prepare a report on the usefulness of OR for the Air Force. In their report, they argued that the Royal Air Force experience had demonstrated that teams of scientists could analyze massive data about the effectiveness of new weapons in operations, drawing out meaningful and significant patterns that could be used effectively by commanders who otherwise would be swamped by unanalyzed data. Leach and Davidson recommended further that lawyers or others adept in presenting information head the teams of civilian analysts, and that these teams have direct access to unit commanders. By the end of the war, every one of the major operational Air Force units had operations analysis teams. Air Force commanders could not help being impressed by the improvement in bombing accuracy, which they called "nothing short of spectacular" and which they attributed to the analyses of OA teams. In the Eighth Air Force, for instance, the claim for bombing accuracy rose from 15 percent in 1943 to 60 percent in 1945 (percentage of bombs dropped within 1,000 feet of the aiming point).[24]

On the other hand, experienced military commanders sometimes rejected the scientists' rational advice. One operations analysis group, for example, recommended that, to facilitate aerial photography of target damage, bomber pilots, after releasing their bombs, continue to fly on course—rather than take evasive tactics. The OA group justified the expected loss of pilots and bombers on the assumption that the information the survivors gathered would save lives on a compensatory scale in future missions.

After the war Morse and other OR enthusiasts offered their services as consultants to industry and academia, encouraging the hiring of a professional staff trained in the new field. They organized professional societies, published journals, and wrote articles and books about a complex of new ideas and methodologies associated with operations research. In 1951, Morse and George Kimball, a Columbia University chemist who had worked with him in OR during the war, published *Methods of Operations Research,* a declassified volume of their summary, compiled for the Navy, of wartime OR activities. The book became a major reference in the field.[25]

But OR enthusiasts initially encountered substantial opposition. Scholars in long-entrenched disciplines with accepted methodologies often found the new ideas and methods had too many rough edges and thin spots as compared to older, well-honed ones. Professionals in these fields sometimes felt threatened—even deskilled—by the new approach. They asserted that it provided merely a variation on well-established methods.

In industry, Morse and Kimball encountered substantial resistance to their endeavor to establish in-house OR units, in contrast to using OR consultants. The would-be innovators concluded:

> Administrators in general, even the high brass, have resigned themselves to letting the physical scientist putter around with odd ideas and carry out impractical experiments, as long as things experimented with are solutions or alloys or neutrons or cosmic rays. But when one or more of them [*sic*] start prying into the workings of his own smoothly running organization, asking him and others embarrassing questions not related to the problems he wants them to solve, then there's hell to pay.[26]

Frustrated in one direction, Morse would take another tack. Experience had taught him that one should have ten endeavors under way if one wanted to succeed in at least one. Knowing that establishing professional organizations has long been a means to cultivate a new field, in 1952 he became the first president of the newly founded Operations Research Society of America. Seventy persons assembled at Arden House, a Columbia University mansion overlooking the Hudson River, to found the society, half of them from the military and the remainder from industry and the universities. The membership of the organization reached about a thousand by 1960, then spurted to 7,500 a decade later.[27]

In their OR campaign, Morse and other OR proponents also tried to infiltrate long-established science and social science organizations:

> The Social Science Research Council gave me a rough time when I told them I believed that operations research differed both in techniques and in subject matter from the social sciences on the one hand and economics on the other and thus had standing as a separate subject. I seemed to convince a few of them. Other O/R veterans also were working for the same ends. We managed to penetrate the sanctum of the National Academy of Sciences or, rather, of its operating subsidiary, the National Research Council (NRC).[28]

Morse and other OR enthusiasts sought to embed it not only in industrial and professional organizations but in academic ones as well. They had some success in engineering schools, especially in departments of industrial management, where the engineering faculty felt at home with the experimental, observational, quantitative, and analytical character of OR methodology. The industrial engineering department at Johns Hopkins University became a leading advocate and disseminator of operations research studies. The Army Operations Research Office (ORO), located near the university campus, had a close affiliation with the Hopkins industrial engineering department. Civilian staff members of ORO had academic appointments at the university and taught operations research courses. The semiautonomous Applied Physics Laboratory, administered by the university, also located nearby, cultivated research relations with the university. The laboratory had an international reputation for its systems approach to projects, including the development of the proximity fuse during World War II.

Professor Robert Roy, head of industrial engineering and later dean of engineering at Johns Hopkins, encouraged these contacts, taking advantage of grants from ORO to expand his faculty and graduate program. Bell Laboratories also helped fund the expanding Hopkins OR program by sending junior employees to the graduate program and by paying their tuition and salary. The Navy sent officers for graduate study as well. By 1960, the industrial engineering department, newly renamed the department of operations research and industrial engineering, had a mix of engineers, economists, statisticians, accountants, and psychologists among its faculty of ten, demonstrating once again the transdisciplinary nature of the systems approach. Over a span of fifteen to twenty years, the department awarded almost a hundred Ph.D.'s, colonizing many universities, industrial corporations, and government agencies with its graduates. The faculty frequently served as OR consultants to industry and government.[29]

The director of the Johns Hopkins University Hospital—then as now a leading medical institution—hearing informally of the expansion of operations research studies at Hopkins, invited Roy's faculty and graduate students to study management at the hospital and to make recommendations for procedural changes. Associate Professor Charles Flagle, trained as an electrical engineer, applied OR to a number of hospital practices. He found "around every corner" stochastic processes, such as the random nature of patient demand on the hospital system, that responded to OR efforts to increase the effective use of resources and to reduce costs.[30] In 1956, Flagle became director of a newly formed hospital operations research division, which was staffed by a number of graduate students from industrial engineering. By the 1960s, a plurality of Johns Hopkins doctoral candidates in industrial engineering were specializing in the application of OR to the medical field.

Other leading institutions, including Morse's MIT, the University of Pennsylvania, the University of California–Los Angeles,

Northwestern University, University of Michigan, and Case Institute of Technology, also cultivated operations research. Faculty members, some of whom had learned OR during World War II, offered courses; published essays, monographs, and textbooks; won government and industrial research contracts; and served as consultants to industry and government. Sometimes contract and grant money made possible the establishment of an OR research institute staffed by professors released from full-time teaching.

The number of graduate seminars, undergraduate courses, books, and articles on the subject increased sharply in the late 1960s. In 1964, *The Engineering Index,* which tabulates articles by subject matter, had no heading for "systems engineering" and only two pages of citations for articles about operations research; by 1966, there was one page of citations under the rubric of systems engineering and six pages given over to OR. In 1969, the number of pages for systems engineering jumped to eight and those for OR to ten.[31]

In contrast to systems engineering, which is a tool for managing projects, operations research is a tool for analyzing and optimizing deployed systems, or operations under way. Systems analysis, to which we now turn, by contrast gives its practitioners a means for evaluating alternative proposals for future projects. From operations research, systems analysis borrowed holistic, transdisciplinary characteristics and the reliance on natural scientists and scientifically trained engineers and their methods. But as we shall discover, social scientists helped determine the character of systems analysis.

]]]]]] Systems Analysis: The RAND Corporation

In 1948, the U.S. Air Force established the Research and Development Corporation (RAND Corporation), which became the seedbed of systems analysis. Over the next decade or so, the RAND Corporation prepared classified and public reports that substantially enhanced the reputation of operations research, operations analysis, and, especially, systems analysis—all planning and managerial techniques dependent on quantitative and systematic analysis of masses of data.

As we observed earlier, General Henry "Hap" Arnold had the leading role in bringing the Air Force to rely on academic and industrial engineers and scientists for the introduction of new weapons. Edward Bowles, a professor of electrical engineering at MIT, who in 1944 became a special adviser to Arnold on communications, radar, and organizational and operational problems, brought operations research to Arnold's attention. Bowles, the chief engineer of Douglas Aircraft Company, and a fellow Douglas engineer, Franklin R. Collbohm, took part in an OR study of the most efficient use of the new B-29 bomber.[32] The team imaginatively recommended that heavy armor be removed from the B-29 to increase its speed, thereby enabling it to elude attacking Japanese fighters.

As the war drew to a close, Arnold discussed with Donald Douglas, whose company had been the third-ranking producer of aircraft during World War II, the possibility of his company taking on long-range Air Force study projects. Friends who hunted and fished together, Arnold and Douglas drew even closer when Arnold's son, a West Point Military Academy graduate, married Douglas's daughter. Collbohm, who had served on Arnold's OR team, proposed on behalf of Douglas that the company organize a group of civilian scientists to study future weapons development. In response, Arnold enthusiastically provided $10 million left over from his wartime research budget to fund the group's study of intercontinental air warfare, an endeavor soon named Project RAND. Initially the staff of about a hundred persons consisted primarily of physicists, mathematicians, and engineers.

By 1948, the possibility of conflict of interest had become clear. RAND might recommend weapons that Douglas was particularly well suited to manufacture. So RAND spun off from Douglas to become a nonprofit corporation heavily dependent on, but not limited to, Air Force funding. The Ford Foundation provided seed money for the new corporation. With headquarters in Santa

Monica, California, only a short stroll from the Pacific Ocean and an hour's drive from the heart of Los Angeles, RAND employees did research in a campuslike atmosphere.

Collbohm became the head of RAND, quickly winning the respect of his professional staff by minimizing bureaucratic control and fostering an ideal university atmosphere that allowed permanent staff and consultants considerable leeway in their choice of research problems. As a result, RAND attracted many academics who preferred the transdisciplinary, problem-solving approach at RAND to the impervious departmental boundaries common in universities. Collbohm also avoided the tight budgetary controls common in industry. Earlier, when the RAND Project was under the wing of Douglas Aircraft, management had attempted to limit the allocation of chalk that the scientists used with their cherished blackboards.[33]

RAND's highly innovative organizational form and practice became a model for other federal contract research centers established later, such as the MITRE and Aerospace corporations. On the other hand, RAND critics, especially in the 1960s, called it a permanent floating seminar in which the stakes were millions of lives and the players were double-domed Dr. Strangeloves devoid of human feeling. Because of RAND's role in formulating nuclear strategy, the Soviet Union labeled RAND an "academy of death."[34] Another observer compared the advisers and consultants from RAND who moved through the corridors of the Pentagon to "the Jesuits at the courts of Madrid and Vienna three hundred years earlier."[35]

RAND became a major influence in strategic-defense policy making as early as 1952 when Ernst Plesset, head of the physics division; Charles Hitch, head of the economics division; and Bernard Brodie, a social scientist, briefed the Defense Department about the physical characteristics of the hydrogen bomb and the broad-ranging impact the bomb would have on foreign policy and military strategy. Giving the first systematic analysis of the subject, Hitch, who served a term as president of the Operations Research Society of America, and his associates dramatically demonstrated the effectiveness of the RAND transdisciplinary approach, especially the use of quantitative analysis, such as game theory, which John von Neumann brought to RAND as a consultant. Not long after the Plesset, Hitch, and Brodie study, RAND researchers prepared another highly influential report on intercontinental ballistic missiles that, as we have seen, shaped the recommendations of the Atlas missile committee headed by von Neumann.

Another seminal RAND report that lastingly shaped its analytical and broad systems approach to problems and issues came in 1954 from a study team headed by Albert Wohlstetter, a RAND mathematical logician.[36] After finding that congressmen possessed strong and conflicting opinions about where the Air Force should locate its bases, the Air Force asked RAND for an objective systems analysis of the problem. Wohlstetter and his associates, seeing their primary task as making the right policy assumptions, took a broad systematic approach. They asked themselves: "What kind of future war are we talking about?" "What are the implications of nuclear weapons technology on base location?" "How does base location relate to the broad mission of the Air Force?"

They assumed that the primary objective of base location lay in insuring that U.S. air forces would survive a nuclear attack from the Soviet Union and then be able to strike back with their own nuclear forces. This assumption became the essence of the strategy of deterrence for which RAND and, in time, the Eisenhower administration became known. This strategy offered an alternative to a first nuclear strike in response to a Soviet Union–launched ground war in Europe, a policy likely to bring "holocaust or humiliation."[37] Instead, the holistic Wohlstetter approach took into account a host of technical factors, including the physical characteristics of aircraft and intercontinental ballistic missiles, and of nontechnical ones, such as the attitude toward the United States of countries on whose territory the bases might be located. By the mid-1950s, researchers at RAND began calling their global, systems-oriented, transdisciplinary approach "systems analysis." They believed that their systems analysis approach encompassed more factors than the earlier operations research approach, although they drew heavily upon it.

Mathematicians, engineers, scientists, and economists all interacted as members of RAND study teams in the 1950s. The economists had a disproportionate influence because their intellectual
horizons embraced strategic questions and at the same time they
could speak quantitatively about costs and benefits.[38] The availability of mainframe computers at RAND in the 1950s enhanced the
economists' analytical power to deal with complexity. The engineers
and scientists proved especially effective in doing conceptual and
feasibility studies of nuclear weapons and satellites, but they left
detailed engineering designs and deployment studies to consultants
from the aerospace industry and academia.[39]

RAND's notable successes with transdisciplinary research
stimulated other organizations, including universities, to embrace
this approach as well. Unfortunately, would-be imitators often did
not grasp some of the subtleties of this method. Even though the
RAND professional staff organized itself by disciplines—which
gave economists, for instance, peer associations and competition
inside and outside of RAND—the staff's problem-solving assignments cut across the disciplinary boundaries in a matrix fashion, an
organizational arrangement we have encountered in the SAGE and
Atlas projects. The economists proved especially adept transdisci-
plinarians because they came from a tradition of equilibrium economics, which emphasizes the interaction of countless factors in
processes seeking equilibria. Mathematics provided a common language for interdisciplinary ventures.

There are other explanations for RAND's transdisciplinary
successes. Wohlstetter, who eloquently spoke the quantitative language of the economists, also reasoned with comparable authority
in other fields. He proved to be a team leader and role model. He
supplied the connective tissue for transdisciplinary efforts, thereby
cultivating "a plant that is very difficult to grow."[40] Hitch, as head
of the economics section, reinforced transdisciplinarity by assigning priority, prestige, and resources to RAND economists according to the difficulty and policy significance of the problems on
which they were working rather than according to seniority.

Unlike university researchers inclined to choose problems bounded by a disciplinary frame, the RAND researchers had to respond to the discipline-transcending nature of strategic policy problems. "If you got serious about the problem," Alain Enthoven, a RAND economist, recalls, "you had to bring in all these disciplines. . . . Focused on the problem, one pursued it wherever it led."[41] His explanation reminds us that participants in SAGE summer studies also abandoned their disciplinary orientations as they tried to solve new problems for which old expertise proved inappropriate. Decades earlier, Henry Ford had said he did not want experts working on new engineering problems because they were expert about old solutions. Persons conditioned by their disciplinary associations tend to see themselves as experts, but few persons working on the front edge of research and development can claim to be expert about problems without a history.

Enthoven gives the Air Force credit for providing RAND researchers the "space" to investigate fundamental problems rather than insisting that RAND become an "answer center" responding to short-range problems. RAND saw problems in terms of years of research, not weeks or months. If the Air Force had not followed such a policy, it is doubtful, Enthoven argues, that RAND could have attracted the highly regarded and competent scholars who came to dominate it in its golden years.

The composition of its board of trustees reveals much about the character of RAND as well as of the West Coast research and development establishment. H. Rowan Gaither became president of the Ford Foundation in 1953, but continued to serve as RAND board chairman until his death in 1961, with the exception of one year when William Webster, president of the New England Electric System, took over. Other board members included William R. Hewlett of Hewlett-Packard, whose headquarters were in Palo Alto near Stanford University; Edwin E. Huddleson, Jr., of the San Francisco law firm of Cooley, Crowley, Gaither, Godward, Castro & Huddleson; Edwin M. McMillan, director of the University of California–administered Lawrence Livermore Radiation Laboratory;

James A. Perkins, president of Rice University; David A. Shepard, executive vice president of the Standard Oil Company of New Jersey; and Julius Stratton, president of MIT. When this board met, the military-industrial-university complex was in session.

]]]]]] Systems Analysis: From RAND to DOD

The success of systems analysis under the auspices of the RAND Corporation made its transfer to other organizations irresistible. The most dramatic illustration of its application was by Robert McNamara who, as secretary of defense during the administrations of John Kennedy and Lyndon Johnson, revolutionized management practices in the Defense Department, no insignificant achievement in an agency with an annual budget of more than $40 billion. McNamara's managerial innovations so impressed President Johnson that he ordered that they be introduced into other government agencies.[42]

After receiving a master's degree at the Harvard Business School and teaching there briefly, McNamara and several of his business school colleagues introduced statistical control, a managerial technique they had acquired at Harvard, into the Army Air Force. Commissioned as officers, they applied their techniques to the operations of the Eighth Air Force, then based in England and operating over Germany. After transfer to India, McNamara rationalized the movement of supplies and personnel from bases in India over the Himalayas to U.S. forces then flying B-29 bombers against Japan. His policies, which emphasized quantitative analysis, closely resemble those associated with operations research.

After the war, McNamara and nine colleagues with similar educational and wartime experiences decided to offer their managerial expertise as a team to a single company. Henry Ford II, who took over the reins of the Ford Motor Company from his grandfather, Henry, hired the young managerial radicals. He hoped that they would revitalize the company.

Soon known as the Quiz Kids and later as Whiz Kids by the established Ford engineers and managers, McNamara and the

Young Turks seemed to be constantly asking questions and usually supplying all the answers as they radically changed management habits and introduced rational cost accounting and decision making.

As McNamara's "greater zeal and self-righteousness set him apart from the other Whiz Kids," Henry Ford II named him company comptroller.[43] With his formidable intellect and impressive powers of concentration and analysis, McNamara did not suffer gladly the older Ford managers and engineers who could not follow his highly quantified reasoning. Some critics said he often used numbers to intimidate rather than to explicate. Others found that he ruled more by fear than by rational technique. Nevertheless, his decisions so greatly impressed Henry Ford II that in 1960 he elevated McNamara to the position of Ford Motor Company president.

His managerial and personal styles soon spread throughout the country by word of mouth and by articles in business journals. He chose to live not in Bloomfield Hills where Ford executives and wives were expected to reside, but in nearby Ann Arbor where he associated with the University of Michigan community. Though a Republican, McNamara supported such liberal causes as the American Civil Liberties Union. His reading matter included religious philosopher Pierre Teilhard de Chardin's *Phenomenon of Man*.

No wonder that John Kennedy's talent scouts thought they had found a Republican with an intellectual style that would impress Kennedy, one who as a cabinet member would bring with him the rational managerial skills for which the president-elect was not known. Kennedy named McNamara head of the Defense Department, with the understanding that McNamara would have a free hand there; that he would not have to join the Washington party circuit even though he became known in intimate circles for his ability to do the dance craze "the twist."

Ensconced in the Pentagon, McNamara determinedly sought control of the enormous, sprawling budget. Not long after entering office, he heard a lecture by RAND's Hitch, at the completion of which McNamara banged his fist on the table, saying, "That's exactly what I want." His "mathematical mind lit up like a light being switched on" as he read *The Economics of Defense in the Nuclear Age,* which Hitch had coauthored.[44] The systems analysis techniques advocated by Hitch and his RAND colleagues seemed to summarize and refine the approach to management that McNamara had taken during the war and during his years at Ford. He quickly named Hitch assistant secretary of defense and Defense Department comptroller, and Enthoven deputy assistant secretary for systems analysis.[45] He also established a programming office and a systems analysis office. Hitch and Enthoven then brought into the Pentagon a number of young RAND experts who came to be known, predictably, as the Whiz Kids. They overhauled budget practices by introducing a RAND-style "planning, programming, and budgeting system" (PPBS), a variation on systems and cost analysis.[46]

Observation of Enthoven's activities at RAND and subsequently at McNamara's Defense Department affords an instructive view of RAND's influence in government. A Stanford graduate, Rhodes scholar, and a Ph.D. in economics from MIT, Enthoven joined RAND in 1956, where he remained until 1960, gaining a reputation, along with Daniel Ellsberg, who later became known for his disclosure of the Pentagon papers, as one of the brightest of the young economists. In his final months at RAND before he moved to the Defense Department, Enthoven wrote a report on America's strategic air forces, the vulnerability of which he considered the major danger facing the western world. The introduction of ICBMs into the world arsenal especially alarmed him. He concluded: "Fundamentally I do believe that the situation borders on the hopeless and the tragic." When his report and his concerns were all but ignored by RAND management, he became disillusioned, even embittered, by the RAND style. A devout Catholic, Enthoven lamented that RAND experts had become resigned to making forty-five-minute slick briefings on subjects of deep consequence and complexity.[47]

Not long after taking his post at the Pentagon, McNamara asked Enthoven: "How many more air transports should the Defense Department be ordering?" Enthoven's thorough systems analysis culminated in the development of a new transport, the C-5A.[48] McNamara found Enthoven an able aide, discussing problems with him almost every day. Only thirty years old when he took charge of the Systems Analysis Office at the Pentagon, Enthoven had the systems analyst's "obsessive love for numbers, equations, calculations, along with a certain arrogance that his calculations could reveal the truth."[49] As he reshaped the entire defense budget according to mission categories, Enthoven's arrogance reportedly drove some Air Force generals mad. Accused of being inexperienced in war, Enthoven retorted to one officer, "General, I have fought just as many nuclear wars as you have."[50]

Despite his sharp rejoinders and supreme confidence in his own methodology, Enthoven appears to have been a persuasive, even eloquent, advocate of the systems approach. Speaking at the Naval War College in Newport, Rhode Island, in June 1963 before a skeptical, even hostile, audience of military officers, he opened by disarmingly arguing that he would not be showing respect for the gravity of his and their responsibilities if he rambled on in a non-controversial and boring way. So he promised to discuss frankly the "programming systems" approach to weapons systems acquisition and strategic policy.[51]

He described the approach as emphasizing objective, or scientific, analysis rather than as placing exclusive reliance on the experience and judgment of military officers. This, however, did not mean, he insisted, the downgrading of the military or "excessive reliance on young and temporary civilians," as some critics complained. He wanted to draw on, and rationally analyze, the experience of the military, because he felt that testing hypotheses about military matters by experiment in peacetime is of limited value, especially hypotheses about nuclear conflict. While he described his systems approach as relying heavily on an objective and scientific method insofar as possible, he added that systems analysis, the essence of the approach, was yet in an embryonic state, still an art rather than a hard science. He defined the scientific method as "an open, explicit, verifiable self-correcting process" that combined logic and empirical evidence. He reminded his military audiences that systems analysis had originated in operations research, which all of the services had embraced years earlier.

On his controversial dependence on numbers, Enthoven remarked:

> Numbers are a part of our language. Where a quantitative matter is being discussed, the greatest clarity of thought is achieved by using numbers instead of avoiding them, *even when uncertainties are present.* This is not to rule out judgment and insight. Rather, it is to say, that judgments and insights need, like everything else, to be expressed with clarity if they are to be useful.[52]

Enthoven, Hitch, and McNamara introduced their artfully scientific PPBS in alliance with the Bureau of the Budget. Before its introduction, the secretary of defense had customarily allocated budget shares to the heads of the various branches of the military; they then earmarked funds for various weapons projects and other activities in their respective services. As a result, no one knew the total budget—as contrasted with the budget of the individual military services—allocated for specific Defense Department missions, such as continental air defense and strategic retaliatory forces.[53] In contrast, the RAND approach defined overall DOD missions, systematically including costs such as training and maintenance, as well as the obvious hardware items such as airplanes and radar. The RAND approach called for defining the mission, planning the forces needed to fulfill them eight years into the future, and budgeting for these forces.

Taking the systems approach, the Whiz Kids, Enthoven chief among them, surveyed national, military, or strategic objectives, then used systems analysis techniques to evaluate alternative means of achieving these ends, ignoring as far as possible the staked-out claims of the Air Force, Navy, and Army. The approach categorized various weapons systems according to the strategic objective for which they were designed. Weapons such as Atlas became "program elements" in the category of strategic retaliatory forces. If current policy called for targeting with nuclear bombs specified cities, industrial sites, and military installations in the Soviet Union, those practicing systems analysis took into account the alternatives of using bombers, land-based missiles, or submarine-based missiles to destroy these targets, then recommended a mix of delivery systems and weapons that would be most cost-effective.

The proponents of such a systematic approach believed that it eliminated budget "games" played by the military, such as getting "the thin edge of the wedge" approved (budgeting the acquisition of airplanes) and then when they were half built informing the secretary of defense of the additional cost of bases, training, tankers, and so on. Enthoven says he would not have been surprised if the Army had come to him and said that it had made a lamentable mistake by spending its million-dollar shoe allocation on right shoes only, that now it needed another million for the left ones.[54] Such remarks did not endear him to the military.

The efforts of McNamara, Hitch, and Enthoven to introduce the systems approach into government met stiff opposition. Military officers reacted negatively to weapons acquisition innovations that broke the momentum of well-established practices and undermined the skills embedded in them. McNamara's critics resisted his centralization of authority in his office, but "critics who begin by complaining about centralization often seem to end up talking bitterly about systems analysis, cost-effectiveness studies, whiz kids, computers, and other manifestations of a scientific spirit that is said to be overriding 'the human factor.' "[55]

Inevitably, a backlash ensued. Enthoven recalls that officers who were assigned to his office and were won over to his approach often received poor fitness reports from superiors who opposed systems analysis. Long-term congressmen who had close ties with the military often complained about young civilian intellectuals who lacked combat experience usurping military prerogatives. Senator Henry Jackson of Washington State had to take strenuous action to prevent fellow congressmen from abolishing the Systems Analysis Office.[56]

Over time, the resistance of the military services lessened, especially after young officers who had worked in Hitch's or Enthoven's office returned to their home service and instituted some variation of systems analysis and a "planning, programming, and budgeting system." Every former Rhodes scholar in the Navy, save one who was too senior, did a tour of duty in Enthoven's Systems Analysis Office, as did Rhodes scholars from other services, thus bringing prestige to systems analysis. In time, the services informally agreed that such a tour would not count negatively in career records.

Eventually, the formidable quantitatively based arguments made by experienced RAND advisers on military matters provoked the military to cultivate in-house systems analysis. The Navy postgraduate school launched a course in systems analysis; Enthoven gave lectures at the service academies. The spread of systems thinking into the services fostered a debate about weapons systems and budget among the services and the Defense Department. Conducted according to—and disciplined by—the rules of evidence and calculations, the debate, in the opinion of Enthoven, raised the level of policy analysis. Systems analysis did not displace debate, it became a framework for it.[57]

]]]]]] Confluence and Convergence: Coping with Complexity

Practitioners of the systems approach claimed that they had the tools to deal with problems of unprecedented complexity not only in the military but in the civil realm as well.[58] In a period that many cultural critics call an era of complexity, this claim attracted new adherents to the systems cult. Systems approach enthusiasts contended that they could "cope with [the] complexity" of society's social problems. "Issues were so complex," one commentator noted, "that people did not know how to manage them well, let alone set up opportunities for effective participative democracy in dealing with them."[59] Society faced a crisis of control.[60]

Social scientists and public intellectuals defined the baffling social complexity to which the systems approach enthusiasts believed they could respond as a problem involving indeterminacy, fragmentation, pluralism, contingency, ambivalence, and nonlinearity.[61] Ecologists, molecular biologists, computer scientists, and organizational theorists also found themselves in a world of complex systems.[62] Humanists—architects and literary critics among them—see complexity as a defining characteristic of a postmodern industrial world. Robert Venturi, a perceptive architect, wrote in 1966:

> I like complexity and contradiction in architecture. . . . I speak of a complex and contradictory architecture based on the richness and ambiguity of modern experience, including that experience which is inherent in art. Everywhere, except in architecture, complexity and contradiction have been acknowledged, from Gödel's proof of ultimate inconsistency in mathematics to T. S. Eliot's analysis of "difficult" poetry and Joseph Albers' definition of the paradoxical quality of painting.[63]

A sanguine contemporary observed that "in recent years the federal government has coped more successfully with large, complex and interacting problems when using a systems approach."[64] President Johnson shared this opinion. Following a landslide victory over Republican Senator Barry Goldwater in the presidential race of 1964, and with the Democrats having gained thirty-eight seats in the House of Representatives, Johnson felt emboldened to wage campaigns on two fronts—Vietnam and American cities. His Great Society Program delineated plans for a war on poverty and especially on urban slum conditions, the kind of complex social problems he and his advisers considered amenable to the systems approach.

The National League of Cities and the U.S. Conference of Mayors issued a joint statement characterizing urban problems as some of the most complex and pressing of their time, equating them implicitly with the problems of national defense:

> The crisis of the City is in reality a *national* domestic crisis. It affects every segment of our society: urban jobless require welfare

paid for by Federal dollars nationally collected; inadequate urban schools cast upon the nation another generation of people unprepared to make their own way; urban traffic congestion adds millions of dollars to the cost of items on neighborhood store shelves around the country. The crisis of the cities belongs to the nation.

The nation cannot dissolve the urban dilemma by ignoring it or by suppressing its manifestations. The nation must face it as it has faced crises in the past, by: acknowledging its basic causes; recognizing the nation's will and ability to resolve them; accepting that objective as its obligation to current and future generations; and reordering national priorities to assure the commitment of the kind and magnitude of resources required.[65]

Writing in the journal *Nation's Cities,* Simon Ramo, the organizer of systems engineering for the Atlas project, argued that over the last decade or two the systems approach, "a powerful methodology," had been successful in applying science to develop military and space systems. "Now," he continued, "it is beginning to be apparent that the same systems approach used to put missiles in silos and satellites in orbit is also well suited to attacking the 'civil' problems of our rapidly decaying and congested urbanized cities." Believing that social change lagged behind technological change, Ramo predicted that at least ten years would pass before the systems approach in conjunction with advanced technology and science would begin to be used effectively to solve urban problems. Then we will enter "our golden age," he optimistically anticipated.[66]

In 1965, the Democratic Congress established several departments and programs to sponsor renewal and construction projects, among these the Housing and Urban Development Department (HUD). The Demonstration Cities and Metropolitan Area Redevelopment Act of 1966 provided aid for the rebuilding of blighted urban areas. The Johnson administration also sponsored urban mass transportation projects and ones to improve urban air and water quality.

Johnson named Sargent Shriver head of the Task Force on Antipoverty Programs. In close touch with McNamara, Shriver chose as his deputy Adam Yarmolinsky, a special assistant to Secretary McNamara. Shriver also mobilized high officials in the Bureau of the Budget, including Henry Rowen, a RAND alumnus. They urged Johnson to introduce "programming, packaging, and budgeting" (PPB), a variation of the Pentagon's planning, programming, and budgeting system, throughout the federal government.

At an August 1965 news conference, Johnson endorsed PPB. He then issued an executive order requiring government agencies to initiate the system. Budget Bureau Bulletin No. 66-3 subsequently instructed all department and agency heads to "establish an adequate central staff for analysis, planning, programming and budgeting."[67] As a result of this order, the heads of government agencies had the formidable task of overcoming institutional momentum and retraining employees to use PPB.

Critics questioned the assumption that a managerial approach that worked for the Air Force would also do so for civil government agencies. One skeptic compared this assumption to the one made by the cargo cults of the Pacific islands of Melanesia, members of which had enjoyed the material abundance brought by the Air Force presence during WW II. The cargo cults assumed that by building mock airfields, control towers, and airplanes they might be able to bring back the abundance once associated with those artifacts.[68]

Following Johnson's bold initiative, high-level administrators would ideally establish goals, or missions, to transcend existing bureaucratic jurisdictions in much the same way that McNamara had transcended the service-specific programs of the Army, the Navy, and the Air Force. Existing departmental and agency programs and projects could be "packaged" and treated as subsystems within an overarching government system. Combining the interstate highway projects, railway and transit projects, and housing projects into regional projects would be an implementation of planning and packaging. After packaging, the system and its subsystems could be scheduled and coordinated in a manner analogous to the way in which systems engineers had managed the Atlas project. In practice, bureaucratic inertia and skepticism among high-ranking administrators, combined with a lack of management expertise at the level of middle management, produced only limited results in response.[69]

]]]]]] TRW's Civil Programs

Persons outside government who had been responsible for introducing the military to systems engineering also aspired to spread the systems approach throughout the civil sector. Simon Ramo, who organized the Atlas systems engineering program and who later became an officer of Thompson Ramo Wooldridge, Inc. (TRW), encouraged the company to broadly apply systems engineering and systems analysis techniques.[70] Ramo maintained that companies like TRW could help solve pressing urban and environmental problems generally by rationally mobilizing technology and science. The systems approach, he believed, could correct the disjunction between technological progress and social backwardness.[71]

When James P. Brown, an experienced marketing executive, joined TRW, Ramo gained an ally who wished to apply systems engineering and project management to solving governmental and commercial problems. Other executives at TRW supported Ramo and Brown in part because the recession in the aerospace field was adversely affecting the company. Governor Edmund G. "Pat" Brown of California, also responding to the aerospace recession, favored funding California aerospace companies ready to apply systems engineering methods to solving sociotechnical problems. TRW won a state contract amounting to $200,000 to develop for Santa Clara County a regional land-use plan designed to provide for population changes, commercial development, housing needs, transportation, and communications.

Johnson's federal government also turned to TRW, requesting that it provide systems engineering training for local government managers, that it do systems studies of water resource management, and that it investigate the reorganization and development of rail and highway transportation systems. By 1969, a number of TRW contracts, considered small as compared to aerospace projects, amounted, nevertheless, to about $10 million. The TRW Systems Group, the TRW Systems Application Center, and the Civil Systems Center took responsibility for the bulk of these and other systems analysis and systems engineering projects.

But by 1975, TRW began to lose interest in civil ventures. It encountered competition from small firms as well as from ones long established in the housing and transportation fields. The company learned that, compared with military project-managing agencies, federal and local government agencies lacked a focused, innovative spirit. TRW found local governments jealously guarding their political jurisdictions, thus frustrating rational regional planning. Further, the company found itself bogged down in drawn-out public hearings involving, among others, hostile activist groups. One TRW executive lamented that difficulties in dealing with the military were as nothing compared to dealing with the City of Philadelphia.

As a recession caused by an energy crisis deepened, TRW management reduced the company's involvement in civil programs as a means of shedding low-profit investments and of consolidating the diversified company around its core manufacturing and military project-managing activities. The company declared that the systems approach worked well in solving technical—as contrasted with social—problems related to clearly defined objectives. Experience with civil ventures also taught the company that experienced aerospace engineers accustomed to working on military projects did not have the commercial touch needed for competitive ventures.

||||||| Schriever: Swords into Plowshares

Other system builders enthusiastically moved from problems of national defense to those of civil society. Both General Bernard Schriever and Jay Forrester shifted their attention from the military to the urban sphere in the 1960s. After his retirement from the Air Force as head of the Air Research and Development Command, Schriever decided to apply lessons learned from managing massive military projects to critical and complex urban problems.[72]

Intent on recruiting not only military management experience but also private sector resources to aid government in improving urban conditions, Schriever founded Urban, Inc. (later, Urban Systems Associates, Inc.). This corporate consortium announced that it had "the necessary brain power to conduct advanced planning in the form of technical studies and systems analyses covering the entire urban problem." Urban, Inc., was designed to conceptualize and plan—not to construct—urban projects, including housing and transportation. Members of the consortium, however, might subsequently be involved in the planned construction.

The consortium included two companies in construction (Ralph M. Parsons and American Cement), two from aerospace (Lockheed Aircraft and Northrop), one from computers (Control Data), another from electrical manufacturing (Emerson Electric), as well as Schriever's recently formed consulting company—B. A. Schriever Associates (later Schriever & McKee Associates). Designed as a profit-making enterprise, the consortium had as its overriding motivation—and this was probably Schriever's emphasis—"*the potential contribution to the solution of the urban problem.*"[73]

Schriever believed that urban problems resulted in no small part from decades of unplanned population growth and concentration. The rehabilitation, job training, industrial ventures, and other remedial projects that various nonprofit organizations and universities had undertaken were only partially successful, in Schriever's opinion, because of poorly defined goals and weak management that reacted to problems instead of anticipating them. (Bringing in managerial talent from private industry would, he predicted, improve the situation.)

Schriever aimed to take advantage of the commitment of leaders in the aerospace industry and the federal government to "public interest partnerships" between government and industry "to solve problems in the public interest on a profit basis." Joseph A. Califano, an aide to Johnson, said:

> Government alone cannot meet and master the great social problems of our day. It will take the matchless skill and enterprise of

the private sector working closely with Government at all levels. It will take public interest partnerships of a scope we cannot yet perceive.[74]

Government, a consortium prospectus added, should function as a prime mover aware of the needs of the country and as a definer of programs that industry carries out. When not funding programs, government should create an environment promising a fair return on investment by private industry. Schriever and his associates argued, "The creation of an environment conducive to social change is the role of government and it would be presumptuous of industry to attempt to usurp this role rather than provide support for it."[75] This call for government-industry cooperation contrasts with the free-enterprise ideology that sees government as confrontational.

The consortium prospectus deployed a vocabulary signifying the systems approach of the 1960s: interface, coordination, mathematical model, computer model, and feedback loops. To provide a background, or database, for its technical studies, the consortium intended to organize a "systems planning team" of about fifty professionals skilled in data acquisition and analysis who would produce an "urban simulation model." This model would provide the general background for studies of particular cities. Schriever would manage the study projects.

He and his consortium staff also felt confident that they could find a way to deal with the traditional checks and balances of civil government and the disparate political jurisdictions that frustrated centralized planning and management.[76] Even though contracts were expected directly from state and local governments, Schriever hoped that overarching funding by the federal government under the Great Society Program would permit planning that could transcend local governments and embrace large geographical areas suitable for systematic planning.

Despite Schriever's leadership, engaging personality, and persuasive rhetoric, and despite having proposed to state and city governments a number of study and planning contracts, Urban Systems Associates failed to obtain a single contract during the fourteen months of its existence from January 1968 to March 1969. The State of Arizona, for example, initially delayed responding to a proposal for a "Program Concept for Central Arizona Regional Development," explaining that nothing could be done until after the next Arizona elections—but later simply turned down the proposal. The consortium saw this as a "lack of initiative and motivation" on the part of the state "to change the 'status quo.' "[77]

After Governor Nelson Rockefeller established the New York State Urban Development Corporation to undertake large-scale urban renewal and industrial development projects, the consortium offered to plan "job training, industrial development, and proper consideration of the environment." The organization turned out to be interested in consortium-member firms locating manufacturing plants in New York State, but it had no interest in the systematic planning studies. Efforts to obtain contracts in other states, including Connecticut, California, Florida, Arkansas, Tennessee, Nevada, Minnesota, and Colorado proved fruitless.

Taking into account the federal government's plan to disperse concentrated urban population into rural areas, Schriever and his associates approached the Department of Housing and Urban Development and the Appalachian Regional Commission, offering to do systematic background studies for the development of new rural communities that would use a systems approach to integrate housing and industrial development. Despite favorable reactions from HUD officials, study and planning contracts did not follow.

Similar frustrations elsewhere led Schriever and his associates to conclude that, in general, ill-defined state and local government structures, lack of clearly understood goals among government officials, general resistance to innovation, and the political pressures shaping economic policy doomed their endeavor. After some months of frustration, the consortium conceded that the civil market required close attention to "the 'democratic' process of 'politicking,' " an activity seen as highly irrational by managers accustomed to giving priority to economic and technical rather than political priorities.

The consortium discovered similar fragmentation on the federal level. The Department of Housing and Urban Development, the Department of Commerce, the Department of Agriculture, the Department of the Interior, the Federal Aviation Administration, the Small Business Administration, and the General Services Administration could not be coordinated for a regional plan in the way that Schriever and the Western Development Division had transformed major industrial corporations into cooperating subcontractors in a missiles project. "In essence," consortium reports observed, "government is not structured along systems management lines where in the simplest case, a program manager for a specific project controls financial resources and technological support, and is responsible for the performance, schedule and financial expenditures." Furthermore, political jurisdiction and geographical regions did not coincide. The piecemeal approach in government, Schriever lamented, had often resulted in "the creation of more problems than have been solved."[78]

He and his associates also found that HUD, which they had expected to fund their various studies and related projects, was not a product-development, product-procuring, or contractor-management agency like the Department of Defense. HUD faced overwhelming problems in attempting to coordinate and finance highly fragmented, politicized regional, state, and urban agencies. As a result, a consortium report concluded, "HUD is not yet a qualified 'customer' for industry. . . . therefore, systems technology-oriented, product-producing industry does not at this time have a government customer for systems management services."[79]

On 6 February 1969, the directors resolved that the consortium should be dissolved. Contrary to Schriever's confident prediction that his experiences in managing military-industry-university projects prepared him for dealing with civil government-industry projects, he found himself confounded as a planner facing an unmanageable tangle of competing political jurisdictions. Having become accustomed in the military to different services and commands being able to submerge their differences when persuaded by superiors that a compelling national emergency existed, Schriever found that civil governments and departments in the 1960s did not find urban problems sufficiently compelling to give up their zealously guarded prerogatives. Schriever found that one American city had 130 public service departments, including water, sewage, transportation, housing, and so on. Systems science and the art of politics proved incompatible.

]]]]]] Forrester: System Dynamics

Like Schriever, Jay Forrester applied his experience with the systems approach to solving nondefense problems, including urban ones. When we last encountered Forrester, he was leaving the Lincoln Laboratory and the Whirlwind project to take a position at MIT's Sloan School of Management, established in 1956. We might reasonably ask if the primary motive for his move was a wish to apply technology to the peaceful arts rather than the martial ones. Forrester, however, succinctly explains his move pragmatically. "The pioneer days of computers had nearly passed," he now argues, adding that "the multiple by which computers improved in speed, reliability, and storage capacity in the decade from 1946 to 1956 had been greater than the multiple in any decade since."[80] If the pioneer days were passing, if the frontier was moving elsewhere, the man who grew up among Nebraska pioneers might well believe it time to move on, too.

His appointment as Sloan School professor became his first professorial position.[81] The Sloan School appealed to Forrester in part because the founders of the school expected it to develop a style different from management schools elsewhere because of its physical and intellectual location in a technical university. This style suited Forrester because it would allow him to utilize his experience in managing the hundreds of engineers and technicians in Division 6 of Lincoln Laboratory.

MIT gave Forrester a year free from designated responsibilities so that he could decide upon his role at Sloan. He considered cultivating operations research, but concluded that while it was interesting and no doubt useful, it did not make the difference between corporate success and failure. "Operations research did not have that compelling practical importance that had always characterized my work," he argues.[82] So he launched a program to introduce system dynamics, a feedback-system perspective for design of industrial, urban, and global systems. He distinguishes between system dynamics and operations research. The applier of system dynamics is compared to an airplane designer, while one doing operations research is like an airplane pilot. Operations research, he continues, is for the decision-making manager, while system dynamics is for the designer of corporate policies. Just as an engineer designs a physical system for desired performance, a manager or political leader can aspire to design policies for social systems.[83]

Forrester's particular genius reveals itself in the complex feedback relationships he finds in all dynamic systems—defense, industrial, urban, and world. In the case of the SAGE system, information about the actual positions of aircraft and missiles was fed back into the computer and compared with the predicted positions. When errors were found, adjustments were made. In the case of system dynamics, when a feedback loop reveals that an urban project is not moving toward the goal that planners or decision makers intend, they change the dynamic behavior of components in the system.

He provides a graphic example of "first-order, negative-feedback." If, for example, a person is warming her hands before a stove, she has a level of warmth as her goal. If she moves the hands too close, information is fed back from the state of her hands that she is erroneously making her hands too hot, so she commands a withdrawal. If she draws back too far, another error signal, or message, is generated that causes her to move her hands nearer to the stove. The hand-warming feedback system is called a negative-feedback system.

"Feedback processes," Forrester avers, "govern all growth, fluctuation, and decay. Feedback loops form the essential structures of change in real systems, and are the organizing structure around which system dynamic models are constructed." Feedback became a topic of conversation at cocktail parties and in corporate board-rooms, Forrester recalls, but he knew that this did not mean that the concept was understood. Superficial conversations about feedback usually focus on simple negative-feedback structures, not on the complex nest of circular and interlocking structures "wherein an action can induce not only correction but also fluctuation, counterpressures, and even accentuation of the forces that produced the original symptom of distress" in a system.[84]

To explain Forrester's deep and intuitive understanding of the systems approach, and feedback, we should recall his experience during World War II when he designed gunfire-control devices. Fire-control devices had hydraulic, electrical, and mechanical components, often incorporating mechanical analog computers as well. Gunfire-control devices sustained a number of feedback loops for transmitting information about the state of the target and the state of the exploding shells back to the fire-control system so that it could correct errors or misses by redirecting the guns. Forrester, like so many other inventors, conceptualized via visual metaphors, but he drew images from systems, circuits, and feedback, not from machines as had been common with inventors in the past. Other prominent pioneers who introduced the systems approach, such as Dean Wooldridge, and Norbert Wiener, the author of *Cybernetics; or, Control and Communication in the Animal and the Machine* (1948), also had had wartime experiences designing fire-control systems, for these embodied the most recent concepts about feedback controls.

Forrester believed that if he used his extensive experience in designing gunfire-control systems as well as command and control systems, and enriched these with his growing understanding of industrial and urban systems, he could delineate the essential structure of social systems. He simulated social systems by computer modeling, portraying them as sets of interacting equations and feedback loops. He applied these generalized computer models to specific social systems by supplying quantitative information specific for the system under study. He took into account political and other qualitative factors by assigning to them quantitative values.

His dynamic-systems approach emphasizes information flows. He envisions information, as well as materials, manpower, and money, as "flowing" through a system. He believes that information flowing through a computer model replicates the information flowing through the actual system being modeled if one relies on extensive interviews with persons involved directly in the system being modeled. Such tacit knowledge, Forrester insists, is far more plentiful and influential than the printed word and number.

After becoming a professor, he found himself by chance discussing with several people from General Electric their inability to adjust stock inventories in relation to shipments. Because of this imbalance and the concomitant irregular production, the hiring and layoffs of workers became chaotic. Forrester recalls:

> . . . I started to do some simulation. The analysis based on the feedback viewpoint from my earlier experience used very simple simulations with pencil and paper on a notebook page. The computation started at the top with columns for inventory, employees, production rate, backlog, and orders. Given these initial conditions and the policies being followed in manufacturing, one could enter how many people would be hired in the following week. . . . That first inventory control system with pencil and paper simulation was the beginning of system dynamics.[85]

Pleased by the ease with which he could help solve a troublesome problem for GE, Forrester made a long-term study of industrial processes and common problems encountered as companies expanded. The young but experienced managers who returned as Sloan Fellows to take his seminars helped him identify such problems and to analyze company structures. He applied system dynamics to problems involving more subtle and less tangible considerations than those associated with inventory levels and their physical variables. In 1961, he described his approach in the book *Industrial Dynamics.* He acknowledged that the industrial dynamics approach drew heavily upon his military systems research into information processing and control that used digital computers.

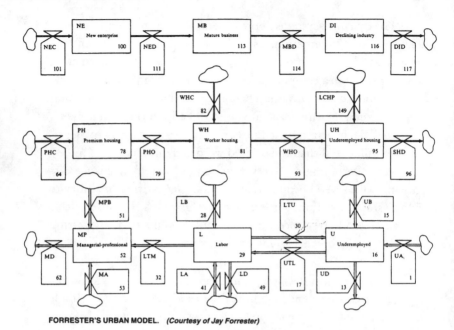

FORRESTER'S URBAN MODEL. *(Courtesy of Jay Forrester)*

He added that the primary source of his industrial simulation models was the design of air defense systems.[86]

Forrester next turned to urban systems. His decision to concentrate on the subject came by happenstance. In 1968, MIT invited John F. Collins, mayor of Boston from 1960 to 1967 and past president of the National League of Cities, to take up residence; Forrester and Collins found themselves in adjoining offices. Their mutual desire to solve outstanding urban problems "led to a joint endeavor to structure and model the dynamics of urban decay and revival."[87] "I suggested to Collins," Forrester recollects, "that we might combine our efforts, taking his extensive practical experience in cities and my background in modeling, and look for interesting new behavioral insights about cities."[88]

Forrester asked Collins to establish a discussion group to meet weekly over a six-month period to seek "insights into those urban processes that could explain stagnation and unemployment." Collins "delivered" the people and they engaged in discussions from which

Forrester's next book, *Urban Dynamics,* evolved.[89] To design a computer model of an urban system, Forrester and his associates defined the boundary between an urban area and the surrounding region and then identified the essential interacting and changing components within the bounded area. The components of particular interest to Forrester included, for instance, the rate of housing construction and the rate of underemployment.

Forrester believed that if he and his associates carefully delineated the boundaries of the urban area and designed the model acutely, they could demonstrate the intrinsic growth and stability characteristics observable in an actual urban area. Aware of the deterioration of the urban scene in Boston and other major American cities in the late 1960s, Forrester declared that a model could "show how an area develops from empty land and eventually fills that land with decaying housing and declining industry to produce economic stagnation. . . . We are interested in the policy changes that can produce revival and a return to economic health."[90] But designing a good model with nonlinear feedback relationships, Forrester insists, is difficult, "akin to the process of invention."[91]

Forrester's urban models portray complex systems not easily comprehended by the nonspecialist. Typically, they have three major "subsystems" (business, housing, and population), nine "levels," and twenty-two "rates of flow." The interaction among these can best be seen by reference to a diagram from *Urban Dynamics* (see opposite page). We see flowing into or out of the level, or pool, of the underemployed (symbolized by a rectangle), five rates of flow. The flows consist of digital units representing birth rate among the underemployed (UB); rate of arrival of the underemployed into the urban area (UA); rate of departures of the underemployed (UD); rate of change of the ratio of labor (the employed) to the underemployed (LTU); and rate of change of the ratio of the underemployed to labor (the employed) (UTL). In the diagram the symbol for a control valve indicates rates of flow. The cloudlike symbol indicates the environment outside the boundary of the urban area. Forrester conceives of the environment as a limitless population sink from which people can move into or out of the urban area. This movement depends partially on the attractiveness of the urban area as compared to the attractiveness of the environment—the underemployed would move from the environment to the urban area, for instance, if there were more employment or more housing for them there. The flow of unemployed persons and information in his models is analogous to the flow of electrons in a circuit; the levels are analogous to the accumulation of electrons in an electrical capacitor.

Forrester's urban models contain numerous information feedbacks. The level of underemployed, for instance, not only has its input and output flows affected by information feedback to the underemployed persons, but also has its input and output flows affected by information feedback to the underemployed about the level, or availability, of housing for them. The flow of underemployed into a region is affected both by the level of underemployment and the level of available housing. In the urban systems model, feedback depends on information flows to people making rational decisions about movement into or out of an urban area.

The urban model helps decision makers decide on goals for levels of underemployment, low-income housing, and so on, and when they take actions or design projects to achieve these goals, thus affecting rates and levels through feedback of information to the underemployed and other persons.[92] Forrester warns decision makers that intuitive judgments about cause-and-effect relationships may not be effective in complex feedback systems, such as an urban system, with their multiple feedback loops and levels. Complex systems have a multitude of interactions, not simply cause-and-effect relationships. Causes may not be proximate in time and space to effects: a decision to increase the availability of housing, for instance, can affect the level of underemployment years later, not unlike the butterfly/chaos effect. A seemingly more proximate cause, such as the shutting down of factories, may hide the effects of the earlier decision to build more housing. Forrester points out that in complex feedback systems, apparent causes may in fact be coincident interactions.[93]

Forrester commonsensically stresses that his models are best constructed by those possessing direct experience with the activities to be modeled. He also believes that managers should take part in the model making because this requires that, in writing precise algebraic equations describing these interactions, they think through the multifarious interactions of complex systems. Forrester eschews the elegant mathematics of many operations research models, favoring, instead, comparatively messy models. "Nonlinear phenomena," he observes, "are central to behavior in complex systems, but nonlinear dynamics lies mostly beyond the reach of formal [elegant] mathematics. . . ."[94] He commends the descriptive case-history approach used at the Harvard Business School because this provides a rich and complex database that can be used in a model. The human mind, he adds, is not able to deal with the inherent dynamic complexity of the data found in an information-rich case history; such complexity requires a computer model.[95]

Forrester recalls that his models suggested that current urban policies lay "somewhere between neutral and highly detrimental" and that "*Urban Dynamics* was the first of my modeling work that produced strong, emotional reactions."[96] Subsequently, his book *World Dynamics* generated substantial criticism as well, especially when it became the intellectual model for the controversial study *The Limits of Growth*; this book predicted a catastrophic economic and social collapse on a world scale unless population and increased raw material consumption were reined in.[97] Economist Carl Kaysen maintained that this system dynamic model grossly simplified economic realities, a charge leveled against system dynamics in general.[98]

Reflecting some years after writing *Urban Dynamics*, Forrester articulated more fully his conviction that his system dynamics approach incorporated the best of the engineering and applied science style:

Before moving to management and social systems, I had an extended career in applied science and its application to engineering, military feedback control systems, computers, and air defense command and control. From that background I see the philosophy of the engineer and scientist as similar to that of the system dynamicist, but as quite different from the philosophy guiding much work in the social sciences. Those working with physical systems gather experience, filter observation through available theory, hypothesize designs, test the components of either equipment or theories, invent, assemble, field test, and redesign. Engineering systems are designed from the inside outward, that is, from components into a functioning whole. Behavior of a system is a consequence of interaction of its parts, parts that themselves must be understood and interconnected. The success of applied science and engineering is measured not so much by published papers as by working devices that do useful things. Engineering is intellectually dangerous. One works beyond the edge of reliable information: one never has resources to make all the tests one would like; a deadline exists; a budget must be met; and prudent risk taking is the setting for every decision.[99]

Forrester personifies in many ways the outstanding characteristics of the engineers of the immediate post–World War II decades. He considered engineering an applied science, used mathematical and graphic models to conceptualize complex artifacts and systems, developed technology pragmatically, and considered the engineering approach applicable to management. Even though he knew from experience that engineering and management involved messy political and social factors, he tended to abstract essential logical structures from complex processes in an effort to comprehend the technical and economic heart of the matter. In so doing, he sometimes failed to take into account the irrational forces and behavior which so often shape policy and practice. As a result, his system dynamics approach, which has influenced countless managers and engineers, has been sharply criticized by social scientists as not embracing real-world complexity. In response, Forrester cites his dependence on rich case histories and experience in conceptualizing his models.

]]]]]] Urban Aerospace Projects

Other persons and organizations experienced with large military-funded projects, especially in the field of aerospace, also offered consulting services and products to regional and urban governments. As one government official put it somewhat uncharitably, aerospace companies emerging from the euphoria of developing complex weapons systems and sending men to the moon promised "Buck Rogers solutions" to the problems of the nation's cities.[100]

In 1964, the State of California awarded four aerospace firms systems analysis contracts for background studies of large-scale problems then facing the state, studies that might lead to remedial projects. California also had the consonant objective of creating systems analysis jobs in the public sector at a time when the aerospace industry faced large layoffs of scientific and technical personnel. This, the state hoped, would prevent the departure of skilled engineers and labor from the state. California aerospace firms, notably those having developed a systems approach, had touted their ability to transfer the systems approach to urban problems. An advertisement of one aerospace firm asserted:

> Snarled freeways. Foul air. Polluted water. Crime in the streets. Soaring medical costs. Overcrowded, understaffed hospitals. Solving top priority national problems is a [company's name] specialty. For over a decade, our balanced blend of systems engineering services and technological skills have been used on America's space and defense programs. Now we are successfully applying this experience to a variety of Civil Systems problems.[101]

After Governor Edmund G. Brown announced the call for competitive bidding for the four $100,000 contracts, about fifty companies submitted proposals. Despite the small amount of each contract, the companies assumed that there would be larger follow-up contracts. North American Aviation received a transportation study contract; Space-General Corporation, a subsidiary of Aerojet-General, obtained a contract to investigate the possibility of using systems techniques to develop programs to prevent and control crime; the Lockheed Missiles & Space Company secured a contract to design an information service to track occupational trends, incidence of disease, and the correlation between employment opportunities and educational requirements; and Aerojet-General had the contract to assess the usefulness of systems analysis and systems engineering as tools to help solve the California waste management problem, especially the reduction of pollution of land, water, and air.

With nearly $1 million in federal funds available for additional systems analysis, California awarded contracts for larger and more focused studies. Lockheed Missiles & Space received $350,000 to study needs for an information system pertaining to criminal justice; Space-General Corporation received $225,000 to analyze the California welfare system; Aerojet-General obtained $175,000 to study solid-waste management in the Fresno area; and TRW, Inc., the successor company to Ramo-Wooldridge, got $220,000 to gather land-use information in Santa Clara County.

After a six-month study of the paper flow in the state, Lockheed recommended that California, using state and local computational facilities already in place, immediately begin the development of an integrated information system. Aerojet-General recommended a three-year waste-management study to prepare a conceptual design for a centralized state waste-management project to be presided over by a state coordinator. The company emphasized the necessity of developing a computer model to represent the interactions of waste with the natural environment. In addition, it recommended the construction of a "socioeconomic model" for incorporating the legal, political, and organizational constraints that would be placed on the designers of the waste-disposal system.

North American Aviation concluded that systems analysis

offered the appropriate approach to transportation planning and, like Aeroject-General, called for computer models that would take into account the rapidly increasing population of the state, the decrease in available land, and the shifting demographic patterns. North American predicted that a computer-based systems approach would bring savings of time and money, less expensive and more rapid freight deliveries, and the preservation of natural recreation areas. The company boldly recommended an integrated transportation system for the entire state. Space-General Corporation, from its systems engineering and operations analysis study of the prevention and control of crime and delinquency, recommended an information system for linking together various criminal justice agencies. It assumed that they would cooperate in the study of criminal activity, prevention programs, apprehension and processing of offenders, and management and treatment of offenders.

In general, newspapers and the public reacted favorably to the studies of the aerospace companies. Office seekers promised that they would use the systems approach as a "powerful tool of technology," one capable of solving a whole array of complex social problems. The Economist of Britain believed the California experiment "the prelude to a national technological assault to engulf mankind in the teeming tomorrow."[102]

In 1966, California and other states spent about $11 million on systems analysis studies, according to one estimate, with New York State alone expending $4 million. Mayor John Lindsay of New York City contracted with the RAND Corporation to undertake studies of select city services, improve those services, and then move toward a new structure of programming and budgeting that would grow out of the studies. Seventeen cities and regional authorities spent about the same as all of the states in 1966. Baltimore and Philadelphia appropriated more than $2 million each.[103]

A 1970 study, The Conversion of Military-Oriented Research and Development to Civilian Uses, lists eighty substantial "system-related contracts" performed by defense-related organizations, most of which are corporations, but some of which are nonprofit research centers. Aerojet-General, Lockheed Missiles & Space, North American

Aviation, TRW, Inc., and the System Development Corporation figure prominently on the listing.[104] The majority of the contracts pertain to urban and regional information, education, health, transportation, crime, and waste-disposal studies. Because of frustrating political circumstances like those encountered by Schriever and his Urban Systems Associates and because of a growing disillusionment in the late 1960s with the so-called military-industrial complex and its systems approach, most of these studies failed to culminate in projects.

In the 1970s, city officials and aerospace managers and engineers met in a series of urban technology conferences at which the voices of the disillusioned were sometimes heard. At one of these, Dean S. Warren, manager of market planning and research for the missile systems division of Lockheed Missiles & Space Company, after noting a number of successful applications of operations research to urban problems, acknowledged nonetheless how experience with urban affairs had eventually tempered the initial unbridled enthusiasm among aerospace firms for this approach. Especially discouraging for him was the failure of local governments to follow up background studies with larger contracts for the procurement of services, a usual practice of the Department of Defense. Like Schriever, he lamented the absence of government officials with sufficient jurisdiction to encompass and enforce the broad concepts of the systems engineer:

> There is rarely one person [in urban affairs] who is empowered to take a strategic overview of a social problem—a vista necessary to really inspire the systems man. There are thousands, hundreds of thousands of customers. And each has a very small pot of money. The customers are politicians, lawyers, or ex-professors of the humanist variety and are not accustomed to quantification. In fact many are deeply and properly suspicious of it even without an understanding of what we offer. . . . Most companies still pursuing this kind of [unprofitable] work are either crediting the cost to public relations or are infected by something akin to a messiah complex.[105]

Warren added that many aerospace firms had lost interest in addressing social problems, not because the problems were unimportant but because aerospace professionals were ill prepared to solve them. He not only doubted the relevance of the aerospace systems approach to urban problems; he even questioned the freshness of the approach:

> The civil engineering fraternity is generally an inarticulate lot. If they weren't, the construction industry would long ago have squelched the grandiose claims of systems men. Focused authority; feedback information systems; detailed, complex scheduling; flexibility in the face of the unknowns of nature; milestone-oriented rewards—program management didn't start in the aerospace industry. The engineer who bossed the construction of Hoover dam did not have the use of computers as do his successors in both industries, but he bossed a gaggle of technical specialties from different companies, had an immense payroll and severe technical, schedule and financial constraints.[106]

At the second Urban Technology Conference in 1972, Cleveland Mayor Carl B. Stokes added to the chorus of disillusionment when he pointed out that aerospace programs enjoyed clear objectives and long-term political support but that urban programs, such as housing for the poor, rarely had either, much less adequate funding. Consequently, he added, aerospace programs focused on technical problems realize stunning successes, while social programs flounder.[107]

]]]]]] Escape from Complexity: The Counterculture

By the late 1960s, skepticism about the effectiveness of the intricate aerospace-generated systems approach as a response to complex urban problems came from many quarters. Daniel Patrick Moynihan, a Harvard professor, sociologist, and adviser to presidents Kennedy, Johnson, and Nixon, looking back in 1972 on the troubled 1960s, fancied that the public then, in contrast to systems experts, was deeply averse to dealing with complexity—and experts—no matter what the approach. Anthropologists, he observed, found people so uninterested, or unable, to cope with complexity that they limited themselves to a simple numerical system consisting only of "one," "two," and "many." Moynihan, not entirely facetiously, pointed out that social problems involved so many variables that the information-overloaded public—like experimental rats in similar circumstances—became hopelessly confused. "Were the great Whigs of the 19th century able to live with complexity because their information was so spare?" Were those Americans seeking to escape from complexity the "simplificateurs" whose rise and dominion the historian Jakob Burckhardt had anticipated a century earlier? Acknowledging that social scientists rightly recognize that complex social problems call for complex responses, Moynihan nevertheless advised the experts to win the confidence of a harassed public by keeping things comparatively simple. Once the public had been suitably prepared, the experts might try a small dose of complexity, he asserted.[108]

In urging social scientists to acknowledge realistically the low tolerance of people for complexity, Moynihan may have had in mind in particular the counterculture enthusiasts of the late 1960s who disparaged big systems, insisting simply that small was beautiful. For many of them, the *Whole Earth Catalog*,[109] which identified simple tools useful in creating and sustaining rural, counterculture communities, proved a better guide to problem solving than the programmatic approaches of urban experts versed in systematic planning.

Because historians and the media today focus on the anti–Vietnam War and civil rights activism of the counterculture, they usually overlook the deep hostility and strong opposition of leading counterculture leaders toward technological systems and the systems approach. They despised the complex systems created by the engineers, scientists, and managers whom we have encountered in the SAGE and Atlas projects. They associated systems science and systems experts with the Vietnam War. California State University professor Theodore Roszak, who in 1969 published *The Making of a*

Counter-Culture: Reflections on the Technocratic Society and Its Youthful Opposition,[110] maintained that rationalizing, planning, and coordinating of the system of production undermined spiritually sustaining and environmentally benign individual and community values. John McDermott, writing in the increasingly influential *New York Review of Books,*[111] appealed to left-leaning liberals, especially to those in academic communities, to help uncover a masterminded massive system of industrial production and military destruction that was laying waste to Vietnam. Charles Reich, a Yale professor, about the same time published *The Greening of America,* in which he claimed that a technological system running out of control had taken charge "of everything—the natural environment, our minds, our lives."[112] Other rhetorically polished and persuasive voices swelled the chorus of lamentation about the corrosive spread of complex technological systems. Prominent among them were Lewis Mumford (*The Pentagon of Power,* 1970), Jacques Ellul (*The Technological System,* 1977), and Herbert Marcuse (*One-Dimensional Man,* 1964).[113]

Critics have claimed that the systems approach, especially systems analysis, contributed to America's failed military policies during the Vietnam War. Even a quarter century after the war, Tom Peters, an influential business consultant and writer as well as a Vietnam veteran, writes that the Vietnam War Memorial in Washington, D.C., "with its 58,000-entry roll call of the dead, is a tragic monument to analytic hubris." He holds McNamara significantly responsible for the "botched prosecution of the war in Vietnam. . . . His grotesque over-reliance on sterile analytic methods and his blindness to the real, human drama of war and politics was his—and our!—undoing." Despite his dismay about the putative use of the analytical approach during the Vietnam War, Peters feels satisfied that "much of the wretched excess of the analysis-is-everything approach to business has been curbed in recent years. . . . Business schools have belatedly begun to repaint their stripes as well: Witness the burst of courses on creativity, quality, leadership, entrepreneurship and human issues in general."[114]

On the other hand, McNamara and Enthoven contend that McNamara did not use systems analysis to analyze Vietnam War policies and problems. In his book *In Retrospect,* McNamara contends that the Kennedy and Johnson administrations failed to take an orderly, rational approach to the basic questions that underlay Vietnam.[115] Enthoven adds, "The problem was not overmanagement of the war from Washington; it was undermanagement. The problem was not too much analysis; it was too little." The administration, he continues, did not make systematic analyses of the effectiveness and costs of alternative military operations. The Systems Analysis Office in the Pentagon had no role in making strategic policy for the Vietnam War. "We did not have a charter," he states, "to analyze the Vietnam operation." Enthoven and his group continued to focus on the horrendous threat that the Cold War might become a nuclear war. When he asked McNamara how he could help him with Vietnam, "he replied that the best thing I could do was to continue the job I was doing on nuclear strategy and NATO.[116]

]]]]]] Criticism of Experts

Criticism of systems science and its experts also emerged from within the systems-approach community. In the 1970s, Russell Ackoff, a leading figure in the fields of OR and social systems analysis, made a dramatic shift from enthusiastic commitment to rejection of the quantitative operations research approach. Earlier, Ackoff had founded an operations research group with Air Force funding at Case Institute of Technology. The group inaugurated the first American operations research program on the graduate level, offering research services to industry and the government. After moving to the University of Pennsylvania in 1964, he and his associates concentrated on solving problems of industrial production and marketing. But in 1971 a sea change occurred in Ackoff's attitude toward operations research. He began to doubt that the quantitative OR emphasis responded as well to marketing problems as a qualitative one. Too "thing" oriented, OR needed to be more people focused, he felt.[117] Formerly a strong advocate of a scientific method he associated with the "hard," or physical sciences,

Ackoff now found the soft aspect of the social sciences appropriate for solving the problems of industrial management. His disquiet assumed such proportions that he tried to persuade the Penn operations research faculty to replace OR with a new program based on different means and ends. Failing this, he and four like-minded faculty withdrew from the OR program to establish a new graduate program and a related research center dedicated to the study of social systems science.

Ackoff quipped that American operations research "is dead even though it has yet to be buried."[118] Filled with mathematical models and algorithms, the articles and reports of academics without practical experience gave few useful formulations of managerial problems and even fewer solutions to these.[119] These OR academics and their students, in Ackoff's opinion, let their abstractions determine their choice of problems, rather than letting the problems troubling industrial managers be the determiners.

Ackoff places his critique of OR in a broad historical frame. Believing that industrial society had moved from a machine age to a systems age, he argued that society should no longer be conceived of as a host of physical and social mechanisms, but as one of interacting systems. In the machine age, humans assumed that mechanistic, cause-and-effect processes produced change; in the systems age, they should, Ackoff reasons, understand that change results not only from mechanistic cause-and-effect, but at a higher, embracing level, from countless purposeful systems with interacting components. During the machine age, would-be problem solvers used analysis in their efforts to understand and control mechanistic change; in the new age, Ackoff contends, they need a holistic approach.

A world composed of systems, Ackoff believes, changes because of "probabilistic confluences" of circumstances that defy deterministic analysis. Therefore, operations research practitioners should not make mechanistic projections intended to help decision makers adapt to the future; practitioners should, instead, help decision makers define future goals and design systems that will approach them. He insists that most humans do not seek quantifiable ends, that their goals are heavily value laden. Those who advise policy makers in planning and designing a future should, therefore, incorporate a humanistic point of view that stresses values. Ackoff now spoke less of problem solving and more of "managing mess." Johnson's Great Society projects failed, in his opinion, in part because of the systems approaches taken by its operations research and systems analysis experts. Instead of mechanistically analyzing problems, they should have sought to manage "messes," the large, dynamic, complex sets of interacting problems with large value components.[120]

To what extent did the intellectual and moral climate of the countercultural 1960s shape Ackoff's new approach? In declaring the death of operations research, he spoke of the deteriorating quality of life and the realization by more and more people that "*optimization* of all the *quantities* of life does not optimize the *quality* of life [my emphasis]." The accelerating rate of change, in the opinion of many persons, he adds, "is getting us nowhere." Citing Jacques Ellul, a French philosopher, and George Wald, an American scientist, both of whom had large counterculture followings, Ackoff contends that more and more people believe they have lost control of their futures. There is a "diminishing sense of progress toward such ideals as peace of mind, peace on earth, equality of opportunity, individual freedom and privacy, and the elimination of poverty." As a socially concerned problem solver, or future designer, sensitive to the social and political currents flowing around him, Ackoff wanted "those of us engaged in helping others make decisions . . . to bring consideration of quality of life . . . into their deliberations"—which operations has failed to do.[121]

Ida Hoos, a research sociologist at the University of California–Berkeley, had a negative, more precise reaction. In *Systems Analysis in Public Policy* (1972) and a similarly critical pamphlet, *Systems Analysis in Social Policy* (1969), she relentlessly attacks the experts who practiced systems science. She argues that, in the name of scientific "efficiency," the systems approach with its use of computer models ran roughshod over local government, that the designers of the models were politically naive, generally bypassing "the checks and balances that safeguard the democratic process."[122]

Hoos labels the State of California's awarding of systems-approach research contracts to aerospace firms a form of "government-by-contract."[123] The activity, she fears, leads to a shadow government that endangers the democratic process. She asks, "Is our present deference to a new set of experts who think they can identify 'the public interest' with a technique and a computer so very different from the deference of the ancient Greeks to the priests and acolytes who interpreted the word of the oracles . . . ?" "The new idolatry," Hoos continues, "differs from the ancient order by virtue of its strongly 'scientific' flavor. This, of course, permeates the systems approach because of its origins and enhances its acceptance as a nostrum for social ills. The ostentatious use of figures and formulae is intended to convey an impression of mathematical precision. . . . Particularly unassailable are techniques and solutions which harness the powers of the computer."[124]

Hoos derides, for instance, the recommendation of the Lockheed systems study that called for an integrated statewide information system in California, to be headed by a single program director. She considers this an unrealistic plan for "an autonomous Information Empire run by a Czar who could circumvent and bypass traditional checks and balances."[125] Only power-hungry officials and technocrats, Hoos believes, will claim that they are able to standardize information-gathering procedures of various state and local authorities, jealous of their prerogatives and jurisdictions, into such a grand scheme.[126]

In our account of the systems approach in the 1960s, we have frequently encountered system builders unwilling or unable to cope with messy political processes that involve checks and balances and multitudinous jurisdictions. This failure contributed substantially to their failures and loss of influence. In the next chapter, however, we encounter a present-day system builder able to cope with messy complexity. His approach suggests a way to be taken if large technological systems are to flourish in the future.

> > > > > > > > > >

V Coping with Complexity:

Central Artery/Tunnel

Rome wasn't built in a day, if it were, we'd have hired the contractor.

CA/T Billboard

▼ We have observed that engineers, scientists, and managers who proved competent, even expert, in the 1950s as system builders in the context of military-funded projects found coping with the political aspects of 1960s civil projects bafflingly difficult. Their messy complexity proved frustrating for persons accustomed to dealing with well-focused projects with clearly delineated lines of authority. Now we turn to a civil project of surpassing complexity, one involving numerous conflicting interests. The history of Boston's Central Artery/Tunnel Project (CA/T) offers a learning experience for those who will deal with other large-scale civil projects in the future.

CA/T involves not only technical, or engineering, problems but also substantial political, social, and environmental issues. Relatively clear national defense goals and a single agency drove SAGE and Atlas. On the other hand, federal, state, and local governments as well as numerous local-interest groups, including those intent upon protecting ethnic neighborhood integrity and the environment, all have their voices in shaping the Central Artery/Tunnel Project. The physical design of CA/T, essentially a highway project, reflects these interests on various levels, even the obvious one of alignment and layout. CA/T becomes congealed politics. Because the designers of projects like CA/T take into account these many factors, the projects are sometimes called "open systems," in contract to the narrowly focused "closed" ones.

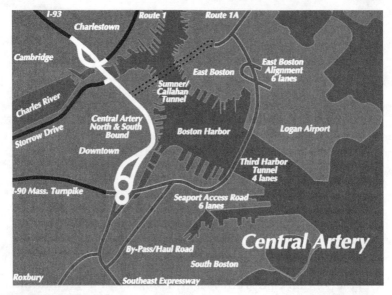

MAP OF THE CENTRAL ARTERY/TUNNEL. THE NEW HIGHWAY SYSTEM LINKS THE EAST-WEST MASSA-
CHUSETTS TURNPIKE (I-90) WITH LOGAN AIRPORT AND ROUTE 1A AND REPLACES THE NORTH-SOUTH
ELEVATED SECTION OF I-93 WITH AN UNDERGROUND EXPRESSWAY. AN INTERCONNECTION JOINS I-90
AND I-93 AT THE SOUTHERN END OF I-93.   *(Courtesy Central Artery/Tunnel Project)*

The leading system builder on this project was Frederick
Salvucci, who differs in many ways from those we have previously
encountered. His style can be compared—and contrasted—with
that of General Bernard Schriever as the Atlas Project system
builder. Schriever and Salvucci both understand how technological
and social systems work. Schriever, with a master's degree in engineering from Stanford, and Salvucci, with a master's in civil engineering from MIT, feel comfortable with engineers in academia and
in industry. Experienced in working within large organizations,
they know how to use political power and at the same time how to
negotiate imaginatively by calling on their own intelligence and
personal magnetism.

Salvucci has had to cope with the messier project. Schriever
extricated his project from a bureaucratic maze (represented by his
spaghetti chart), but Salvucci found bureaucratic spaghetti unavoidable; he eventually developed a taste for it. Schriever found local pol-

itics frustrating when he tried his hand as a planner of civil projects; Salvucci proved to be a master of parochial, ethnic, and interest-group politics in Boston. Schriever had to contend with politicians who wanted to politicize the Atlas Project, but, in an intense Cold War period, he was able to thwart them by appeals to national defense. Salvucci drew on an endless array of justifications and countermoves to outflank politicians and leaders of organizations and groups whose interests ran counter to what he considered the overriding goals of CA/T. Environmental mitigation did not become a part of Schriever's vocabulary; Salvucci will define it for us. Atlas needed a Schriever, Central Artery/Tunnel a Salvucci—although both were system builders, they were not interchangeable.

Expected to cost at least $10 billion, CA/T is the largest public works project under way in the United States, a project comparable in scale, engineering complexity, and social impact to a similar urban project now under way at the Potsdamer Platz in the heart of Berlin, Germany.[1] Scheduled for completion in 2004, the Boston project, involving a 7.5-mile highway construction corridor, or 160 lane miles, will bring to a close the construction of the massive interstate highway network begun during the administration of President Dwight Eisenhower.[2] The Central Artery portion of the project will remove the "green snake," the existing elevated central artery, completed in 1959, which creates an enormous eyesore that separates the North End and waterfront of Boston from the rest of the downtown and which has become clogged with traffic almost three times as heavy as planners originally anticipated. Driven by technical and economic values, the engineers who designed and constructed the green snake left a bitter legacy following their demolition of buildings on thirty-six city blocks. The harbor tunnel portion of the new project will facilitate the movement of traffic to Boston's Logan International Airport, while the underground Central Artery will improve the movement of interstate highway traffic through the Boston region. The project also calls for a new bridge across the Charles River from Boston into Cambridge. The project will involve 3.7 miles of tunnels, 2.3 miles of bridges, and 1.5 miles of surface streets.

THE ELEVATED CENTRAL ARTERY SHOWING CONGESTED TRAFFIC IN 1993. *(Courtesy Central Artery/Tunnel Project)*

Proponents promise that the completed project will variously enhance the existing traffic capacity, separate regional highway traffic from the local, and, after the removal of the elevated highway, provide twenty-seven acres of new open space in downtown Boston. Advocates also predict an increase of 15,000 construction, service, and support jobs during peak construction and, in the long run, hundreds of millions of dollars in increased business sales and personal income in the Boston region. The designers and planners also boast that all of this will be done without the violation of any neighborhood community or the loss of any residential buildings. Supporters celebrate the project as "the largest, most complex highway design and construction project ever undertaken in the United States." In 1991, project proposers optimistically projected a price tag of $5 billion; present pessimistic estimates almost double that amount because of increases in the project's scope and inflation.[3] Those with a sense of history placed the new project in a long tradition of great Boston civil engineering endeavors, including the creation of the great landfill that created Back Bay. The environmentally sensitive project will become, the optimists declare, a model project for other cities to emulate.

]]]]]] Antihighway and Urban Renewal Activists

Both the Central Artery/Tunnel Project, and Salvucci as system builder, have their roots firmly embedded in concerns expressed and actions taken years earlier by an antihighway movement. During the 1960s, a number of socially motivated young persons, including Salvucci, took government positions that would give them some influence over the deployment of federal funds for rapid-transit projects. They believed they could oppose the construction of highway systems that cut through and destroyed neighborhoods and that caused, as well, air and noise pollution. Having seen how rapid-transit projects, especially subways, created construction jobs and provided transportation from inner-city communities to employment sites, they hoped to correlate rapid-transit and anti-poverty efforts. In various parts of the country, and especially in Boston, transportation became a tool for idealist activists seeking to improve the opportunities available to those oppressed by poverty or prejudice.[4]

During this period, the Massachusetts Department of Public Works (DPW; later renamed the Massachusetts Highway Department [MHD]) planned a series of expressways into Boston to be funded mostly by the federal government. An inner belt connecting these expressways would have demolished 3,800 homes, a large number of commercial buildings, and acres of parkland, including some of Boston's "Emerald Necklace" park system. The Inner Belt plan would have eclectically affected Roxbury, a low-income black neighborhood; the Back Bay Fens, a park adjacent to major cultural institutions; the prestigious residential neighborhood of Brookline; and working-class neighborhoods in Somerville, Charlestown, and Cambridge.[5]

Fresh out of MIT in 1962 with his master's degree in civil engineering and a thesis on the economics of urban transportation, Salvucci found a job with the Boston Redevelopment Authority (BRA), the planning agency for the city. He joined other young BRA professionals to discuss, at their regular lunches on the Boston Common, ways to deflect the highway juggernaut and to provide at the same time a transportation system responsive to Boston's transportation problems.

Realizing that the expertise of the highway engineers and lawyers could cow the neighborhood groups, Salvucci and his allies responded positively to the request of a Catholic priest from a Cambridge parish to act as consultants on technical matters in the fight against the Inner Belt. They expanded their resistance front beyond Cambridge by forging coalitions of middle-class whites, blue-collar workers, and working-class blacks. Later, when he became responsible for the formation of coalitions to support CA/T, Salvucci drew on these earlier experiences and allies.

In the late 1960s, Dan Hays, then mayor of Cambridge, convened the Advisory Committee on the Inner Belt that included, besides Salvucci, Mark de Wolfe Howe, an eminent legal scholar; John Kenneth Galbraith, the influential economist; and Daniel Patrick Moynihan, a future U.S. senator. All came from the Harvard faculty. At one meeting, held above a restaurant in Harvard Square, Galbraith persuasively, even passionately, argued that at a time when the city and even the country verged on tearing themselves apart over problems of race, it would be the height of irresponsibility to allow a highway to destroy multiracial communities that could act in unison to save themselves from the highways.[6]

]]]]]] The Study Group

In the 1970 Massachusetts gubernatorial race, both Kevin White, the Democratic candidate, and Republican Frank Sargent opposed aggressive expressway construction within greater Boston.[7] After his election, Sargent convened a blue-ribbon committee, chaired by Alan Altshuler, an MIT professor of political science who specialized in the politics of citizen participation and public transportation, to advise him about outstanding transportation problems. After characterizing Massachusetts highway planning as a "pathological" process, the committee's report stated bluntly, "We perceive a great mindless system charging ahead." Expressways within Route 128 will probably be built, the report concluded, "solely because they involve ten-cent dollars from the state standpoint. . . . The rigidity of federal aid policy sticks you [Sargent], along with other state and local officials throughout the country, with highly unpalatable choices."[8] "Ten-cent dollars" referred to federal highway legislation that provided ninety cents from Washington for every ten cents appropriated locally for highways. The committee recommended a moratorium on highway construction in the Boston area and further study of alternatives. These were bold political recommendations at a time of high unemployment in the construction trades and when the appropriation of a billion dollars in federal highway funds hung in the balance.

Sargent then convened a long-term study group modeled upon the summer study approach we encountered during the formative period of the SAGE project. Representing diverse points of view, the sixty or so multidisciplinary participants (engineers, lawyers, architects, economists, and ecologists) explored the effects of various kinds of transportation on urban, economic, and community development. Besides conducting a research program, the study group heard presentations from interested parties, including highway department officials and antihighway activists. John Wofford, a young lawyer, Rhodes scholar, Harvard graduate, and an expert on transportation law, who headed the administrative staff of the study group, is representative of the young professionals drawn into the highway controversy.

After two years of meeting several evenings a week, the study group, which included Salvucci, produced a report, the Boston Transportation Planning Review. The report recommended against either the construction of the Inner Belt or the extension of major highways into the urban area. As alternatives, it proposed and summarized the pros and cons of a complex set of projects that would place a third tunnel under Boston Harbor, set the elevated Central Artery underground, and extend several Boston transit lines. The highway-building interests with their more direct and reductionist approach to solving transportation problems felt that their point of view had been "overwhelmed by the politics of the situation."[9] They favored a closed rather than an open systems approach.

The National Environmental Policy Act of 1969 also complicated matters. The law required the preparation of an environmental impact statement (EIS) for federally funded transportation projects. An EIS, usually a lengthy document, identifies in detail the various ways in which a proposed project will positively and negatively affect the environment, including air, water, habitats, and parklands. For the proposed transportation projects, the study group prepared one of the first environmental impact statements.

Sargent, advised by the report and by Altshuler, whom he had named secretary of transportation and construction, ruled against the construction of more general-purpose surface highways in the Boston urban core, favoring, instead, investment in urban transit. Sargent also proposed the construction of a managed-traffic, as contrasted with general-purpose, third harbor tunnel as well as the depressing of the elevated Central Artery.[10] Managed-traffic highways and facilities would restrict traffic to multioccupancy and specified commercial vehicles.

In league with other protransit governors and mayors, Sargent brought about a change in federal highway funding legislation. Formerly, states would lose federal funds if they decided not to construct highways for which funds had been appropriated. Under the modified legislation, interstate highway funds could be transferred to a general fund, then the state could request that the amount be appropriated for urban transit.

]]]]]] Salvucci and the Origins of CA/T

In the 1974 gubernatorial race, Michael Dukakis, known for his dislike of highways, defeated Sargent. (Dukakis was relected governor for four-year terms again in 1982 and 1986.) He then named Salvucci secretary of transportation. Salvucci's father was a bricklayer, so, with an inborn empathy with construction craftsmen, the son, who had never felt entirely comfortable in organizing opposition to the expressways, especially when construction workers were unemployed, favored depressing the Central Artery. He saw this as

FRED SALVUCCI (LEFT), DANIEL MCNICHOL OF THE CENTRAL ARTERY/TUNNEL STAFF (CENTER), AND RICHARD JOHN, DIRECTOR OF THE TRANSPORTATION SYSTEMS CENTER, U.S. DEPARTMENT OF TRANSPORTATION. MODEL OF THE CHARLES RIVER BRIDGE IN FOREGROUND. *(Courtesy of Richard John)*

a very special highway that would not raze houses or destroy neighborhoods.

He also supported the construction of the third harbor tunnel, provided it would exit on Logan Airport property rather than in an East Boston neighborhood. Having served as a city hall manager of East Boston and always holding an ear to East Boston ground, Salvucci considered the tunnel project "great, as long as it is out of the neighborhood. . . . It's not a clean position in East Boston but it is respectable enough that I can articulate it in crowds of 200 people and come out alive."[11]

Salvucci's opposition to destructive highway projects plus his commitment to construction projects that created jobs at the same time they respected the integrity of ethnic neighborhoods stemmed from his Boston roots. As a young man, the "nasty" treatment his grandmother received at the hands of the Massachusetts Turnpike Administration had greatly disturbed him. An immigrant widow unable to speak English, she was suddenly ordered by the authority to vacate her house in the Brighton section of Boston within ten days in order to clear the route for an urban highway. She never received fair compensation for her property, much less for the fright and mental anguish that she suffered. Salvucci recalls that his grandmother and the other immigrants had "to crawl on their knees to some politician" to get help in the negotiations over price.[12]

As a student of civil engineering at MIT, he had another experience that deeply influenced his attitude toward large-scale technological projects. A respected civil engineering professor asked Salvucci's class to design a hydroelectric project for an imaginary Latin American dictator in an economically undeveloped country. When Salvucci's team presented a design both massive and handsome, the professor asked, "Why would you let a two-bit dictator lead you to allocate scarce resources to a dam in a country populated by poor peasants so he would have his name on it?" They could answer only that bigger was better.[13] The few seconds the instructor took to make this point became the most memorable event to Salvucci of his undergraduate years.

After Dukakis named him secretary of transportation, Salvucci wanted a central artery and tunnel project as proposed by Sargent's study group, but Dukakis opposed the project. He had an almost irrational dislike of automobiles and found the project inconsistent with his plans for urban transit. One of his political opponents observed:

> To Dukakis, highways and automobiles were really not inanimate objects but animate objects which he hated as you would hate individual evil in people. . . . the only time Michael Dukakis became emotional, at least in my time in public office . . . was when someone discussed transportation.[14]

Yet Salvucci brought Dukakis—and many others—around. Considered Machiavellian by some because of his subtle negotiations, Salvucci possessed intuitive insights into the underlying motivations of key actors.[15] He mustered comprehensive arguments that not only responded to his governor's doubts about the Central Artery/Tunnel package but that also appealed on diverse grounds to a broad coalition of supporters.

His well-publicized prospectus for the Central Artery stressed the reduction of traffic congestion; the more rapid traffic flow to and from Logan Airport; a ceiling on downtown parking-lot construction in order to encourage use of rapid transit—which would be extended; employment for the building trades, contractors, and engineers; and rigorous adherence to environmental law. He insisted that residences would not be taken or neighborhoods violated—no more sad stories of displaced grandmothers.

Salvucci maneuvered so that the third tunnel would not take any buildings or land in East Boston. He and the governor waited patiently until new Dukakis appointees on the Massachusetts Port Authority agreed that the tunnel would surface on airport land. Salvucci promised East Boston residents that not only would the tunnel take none of their homes but that the growth of traffic to and from Logan would be constrained by giving buses and other high-occupancy traffic priority in the use of the tunnel.

Depressing the Central Artery, Salvucci emphasized, would enhance the appearance of Boston by removing the scar that separated the North End, an Italian ethnic enclave, from downtown; depressing the highway would also restore the aesthetic and historical quality of downtown Boston, thereby appealing to the tourist trade; and removing the elevated highway would provide about twenty-seven acres of land that could become downtown parkland. Salvucci's package appealed to a host of interest groups, but of course not to all. The executive director of the Federal Highway Administration (FHWA) believed that the $300 million depression would not solve Boston's transportation problems, that it was a "highway beautification project [for a one-mile stretch of road]."[16]

Salvucci also predicted that the passage of an access road to the other entrance to the tunnel through mostly abandoned, contaminated industrial property in South Boston would help develop this depressed area economically. Why, he asked, should industry move to the Route 128 beltway surrounding Boston, thus taking wetlands and forest, when such an access road would rehabilitate a wasteland and provide jobs for inner-city poor?

In his good ethnic Boston accent, Salvucci appealed to environmentalists, downtown businessmen, professional engineers and architects, ethnic neighborhoods, and the poor. The earlier demands of contractors and building trades for jobs through highway construction had been sidetracked by the creation of jobs through the extension of rapid transit; now Salvucci and Dukakis could promise even more jobs. When Ed King, a political opponent, spoke of renewing the Inner Belt proposal, "his friends among the contractors and building trades responded, 'God bless you, Eddy, but we see the jobs with transit, so let it go.' "[17]

]]]]]] Federal Funding

With a coalition of neighborhood and other activists and interest groups, Salvucci faced the daunting problem of obtaining federal highway funds—the ten-cent dollars. The Massachusetts congressional delegation would have to shepherd through Congress bills that would provide those dollars, but first they needed to poll their constituents on the sensitive transportation issue. For instance, Thomas Philip "Tip" O'Neill, Jr., Speaker of the U.S. House of Representatives from 1977 to 1987, wanted to know where the activists and voters in East Boston stood. His support depended on theirs.

O'Neill's father had organized the bricklayer trades, so the son consistently supported construction—unless it negatively affected the East Boston constituents whom he represented. He opposed early plans for the third tunnel because it would take East Boston homes. (After Salvucci in 1983 changed the alignment so the tunnel surfaced on Logan Airport property, this objection was removed.) Brian Donnelly, a Democrat in the House of Representatives, recalled a meeting that he and Salvucci had in Boston with O'Neill. When Salvucci rolled down a map of the tunnel, O'Neill said, "What tunnel? We're not building any tunnel. . . ." He came around after Salvucci and Donnelly predicted that the "the trade unions are going to be down there marching on you [if you trade in the tunnel]"; and after being assured that Salvucci had "worked the East Boston Community" and that "no homes would be taken," the Speaker responded, "Well, if East Boston is with it, I am with it."[18]

O'Neill and other Massachusetts congressmen then faced the formidable opposition of President Ronald Reagan's administration and the Federal Highway Administration. Both argued that, since CA/T had not been approved before 1981, funding would violate the Transportation Act of 1981, which provided only for funding of interstate highway projects approved before that date. The FHWA also doubted that the artery would provide transportation benefits commensurate with costs; the Reagan administration labeled the project "pork-barrel" spending that would—in conjunction with similar highway projects—add substantially to the federal deficit. On administrative grounds, the FHWA had the power to block construction by arguing that the artery design was not included in the 1981 estimates or by not approving the environmental impact statement required by law for the project. Reagan could veto any congressional bill designed to override the administrative decisions of the FHWA.

To overwhelm the opposition of the FHWA and Reagan, Speaker O'Neill used his power base. If "some flunky" in the Washington bureaucracy tried to block an O'Neill initiative, O'Neill called high-ranking Reagan staff members and told them what he wanted. They would reply, "Tip, it will be through." Reagan would sometimes sign an O'Neill-favored bill that he opposed; then he would "kick the hell out of the Democrats two days later." Such actions did not disturb O'Neill: "Hey, as long as he signs the thing, that's all I'm interested in. When he starts criticizing the thing, I'll call him unfair."[19]

House Public Works Committee Chairman James Howard, a Massachusetts Democrat, told O'Neill, who planned to step down after the 1986 election, "I am going to give this [CA/T] to you as a going away present."[20] Some Republicans then began to speak of "Tip's tunnel." Subsequently, President Reagan observed, in connection with the Boston project and other special-interest highway and transit projects, "I haven't seen this much lard since I handed out blue ribbons at the Iowa State Fair."[21]

With a Republican majority in the Senate to counter the Democratic one in the House, and a Republican president in the White House, organizing the present for Tip proved horrendously difficult. Senate committees opposed authorization of funds for CA/T; the president threatened to veto the transportation bill of 1986, which included funding for CA/T.

Though aware of problems with Congress, Salvucci and other Massachusetts officials concentrated on overcoming the opposition of the Federal Highway Administration and its head, a longtime Reagan loyalist from Texas, Ray Barnhart. Vigorously opposed to CA/T, he impatiently complained that Massachusetts and Boston had "screwed around for so many years in such an antitransportation mode. . . . Coming from Texas, my first inclination," he added eloquently, "was to let the bastards freeze in the dark."[22]

After the Massachusetts delegation argued effectively that Congress had included a portion of CA/T in an interstate cost estimate in 1976, the Federal Highway Administration took another tack by refusing to accept the environmental impact statement prepared for the artery part of the project. Grounds for the rejection included: the presumed limited transportation benefits in comparison with a cost of $1.3 billion; the intended taking of some parkland (broadly defined to include a parking lot owned by a state park agency); serious operational and safety problems; and the displacement of ninety-seven businesses. In 1985, Massachusetts presented its formal response to the FHWA's objections with an analysis purporting to show that the artery depression would have a better ratio of benefits to costs than other major interstate projects supported by the FHWA. These included New York City's Westway and the Century Freeway in Los Angeles.

Barnhart, having then seen traffic congestion in Boston firsthand and taken into account improvements made in the design of the artery, accepted the EIS. This cleared the way for negotiations about funding. Elizabeth Dole, U.S. secretary of transportation, acting for the Reagan administration, "tore me up," Barnhart recalled, for accepting the EIS.[23] Salvucci proceeded to negotiate a compromise on funding: the FHWA agreed to support congressional legislation authorizing federal interstate-completion funding for the harbor tunnel and the north and south portions of the artery; Massachusetts would have to find other federal and non-federal funding elsewhere for the central portion.

The U.S. Department of Transportation and the Office of Management and Budget, on the other hand, attempted to block funding for CA/T and other construction projects by having the president veto the transportation bill. In so doing, the president singled out two projects, CA/T and the Los Angeles subway, as representative of excessive funding for narrow special-interest highway and transit construction projects. Determined to override the veto and show that Congress—not Reagan—dominated domestic policy making, both houses overrode the veto, the Senate by one vote.[24] Democratic senators obtained that one vote by telling newly elected Democratic Senator Terry Sanford of North Carolina that if he supported the veto they might drop their support for tobacco subsidies.[25]

Congressional negotiations and decisions affecting the project were thickly intertwined with other issues before Congress. The Massachusetts congressional delegation and others who pushed the project over a maze of hurdles took into account the interests of, and made promises to, a host of ad hoc allies. These included congressmen who aimed to reassert Congress's primacy in domestic policy making; others who wanted mass transit; some who wished to raise speed limits on rural expressways; others who wanted to continue tobacco subsidies; some who were protecting billboard owners; and still others who were seeking other highway and transit projects from "Altoona, Pennsylvania to Los Angeles, California."[26]

]]]]]] Initiating the Project

With FHWA approval and the federal funding authorized, Salvucci turned to the problems of organizing the management of the project. As the state secretary of transportation, he needed to designate the state agency that would have responsibility for overall management. The Massachusetts Bay Transportation Authority (MBTA) had a better reputation for managing construction projects involving tunneling, especially in the transit field, than did the Massachusetts Department of Public Works. In the post–World War II years, the DPW, whose staff had been reduced by economizing layoffs, relied increasingly on contracting out the design and construction of major projects to private engineering firms.[27]

Salvucci decided nevertheless to assign overall responsibility to the DPW. He reasoned that the state needed a revitalized DPW, and if the management of CA/T were turned over to the MBTA, this would further demoralize the DPW. Furthermore, the FHWA, through which funds would be channeled, preferred dealing with highway, not transit, agencies. Salvucci also decided against having the DPW subcontract management to the MBTA. Wise in the ways of the bureaucratic world, he feared that an unhappy DPW "could sabotage the project. Papers would get lost, that kind of stuff would happen."[28]

To compensate for DPW limitations, Salvucci insisted that it use an exceptionally experienced and capable contractor as project manager, or management consultant. Influenced by Salvucci, the Massachusetts Department of Public Works in 1985 entered into an agreement with Bechtel/Parsons Brinckerhoff (hereafter, Joint Venture), a joint-venture firm, to prepare a preliminary project management plan—on the assumption that federal funding would be forthcoming. The Joint Venture had been formed by Bechtel Civil, Inc., a division of the Bechtel Corporation of San Francisco, and by Parsons Brinckerhoff Quade & Douglas of New York, two of the world's largest consulting and management engineering firms and ones with outstanding records for managing large-scale construction projects.

In August 1986, the Department of Public works issued a "transition contract" to the Joint Venture which agreed, in contractese, to provide "services required to develop a comprehensive work program for establishing the tasks to be performed by a Management Consultant to manage, coordinate, schedule, direct and review the work being prepared by other design consultants for the Third Harbor Tunnel/Depressed Central Artery Project."[29] This one-year contract would be followed by a series of limited-term contracts ("work programs") specifying and extending the services of the Joint Venture as the manager of design and construction of CA/T. By 1993, the twelfth work program was in effect.[30]

Bechtel took 55 percent and Parsons Brinckerhoff Quade & Douglas 45 percent interest in the Joint Venture. Bechtel, strong in the management of both design and construction, and Parsons Brinckerhoff Quade & Douglas, strong in the design of highways and transit facilities, complemented one another. Some who approved of the Joint Venture spoke of a marriage made in heaven. The two firms had joined together earlier as contractors for the San Francisco Bay Area Rapid Transit Project (BART).[31]

Both firms have long and interesting histories. In 1886, William Barclay Parsons founded the Manhattan consulting engineering firm. At the turn of the century, the Parsons firm designed New York City's first full-scale electrified subway, then went on to do a number of hydroelectric, bridge, and highway projects. Parsons, a man of broad experience and talents, published *An American Engineer in China* (1900), which recounted his charting of a thousand miles of railway in China, and *Engineers and Engineering of the Renaissance* (1939), in which he shared his knowledge of Leonardo da Vinci and other Italian architect-engineers.

By 1990, the successor firm, Parsons Brinckerhoff Quade & Douglas, employed more than 2,000 persons. The company also specializes in environmental impact statements and in solving problems associated with air and water quality, noise and vibration abatement, ecology, and land use. It also has experience in providing for public participation in the design of its projects.[32]

Larger than Parsons Brinckerhoff Quade & Douglas, performing work valued at $5.6 billion, and employing about 30,000 persons in 1990, Bechtel usually ranks as one of the four leading U.S. consulting-construction firms. Family-led for four generations, Bechtel achieved a national reputation when it joined with several other prime contractors to build the massive Hoover Dam from

1931 to 1936. Today the Bechtel Group, a holding company, presides over a group of six companies, including ones concentrating in energy, chemicals, and construction. Another Bechtel entity, Bechtel Enterprises, assists clients in obtaining project funding.[33]

Complex, large energy, natural resources, and transportation projects form the core of Bechtel Group activities. It encompassingly handles the multiple facets of large-scale projects, including feasibility studies, financing, project conception, design, procurement, construction, and the management of contractors. The professional characteristics of its employees cover an enormous spectrum, including, for instance, several corporate economists who make worldwide economic and political analyses for use by the firm's strategic planners.[34]

With projects throughout the world, especially in the Middle East and Africa, Bechtel companies have built a Saudi Arabian industrial city for a population of 275,000 persons, in addition to building both the King Fahd and the King Khaled international airports. Among other Bechtel projects have been the English Channel tunnel and a Disney-MGM attraction in Florida. Like a number of other large consulting-engineering enterprises, the Bechtel Group has the holistic competence to finance what it designs and constructs, and then to manage and operate what it has financed, designed, and constructed.

Because of their heavy dependence on government contracts, large engineering-consulting firms have long relied on political connections and influence. Not surprisingly, their executive officers often have well-developed political sensibilities.[35] A wide sector of the public became aware of Bechtel when President Reagan named two Bechtel principals, George Shultz and Caspar Weinberger, as heads, respectively, of the State Department and the Department of Defense. Their presence in the Reagan administration led some critics to suggest that Salvucci wanted Bechtel as project manager to insure Washington's ongoing support for the project.

Bechtel's contacts with political leaders have been so close and its contracts so large that the enterprise has attracted the attention of investigative journalists. In the late 1970s, a small group of reporters in the San Francisco area decided that the public ought to know more about "the most powerful private [family-owned] corporation in the United States and arguably the world."[36] Several journalists associated with the newly formed Center for Investigative Reporting did an "exposé" of Bechtel that they subsequently published in the journal *Mother Jones* in 1978.[37] The essay focused on persons like Shultz and Weinberger who moved through the "revolving door" between industry and government as well as upon John McCone, a Bechtel principal who, as head of the Atomic Energy Commission, got the company "building nukes" and then, as director of the Central Intelligence Agency, reputedly used Bechtel "to provide cover for spooks."[38]

The Nation sharply criticized Bechtel for its close dealings with the Arab states and for distancing itself from Israel. Critics insisted that the company kept its lucrative business with the Arabs by de facto observance of the Arab demand that those with whom they dealt not do business with the Israelis.[39]

]]]]]] Environmental Impact

As one of its early responsibilities, Bechtel/Parsons Brinckerhoff, or the Joint Venture, prepared an environmental impact statement on behalf of the Massachusetts Department of Public Works. State and federal agencies required the department to submit the statement for approval before construction began. SAGE and Atlas engineers had to take into account technical and financial considerations as well as military tactics and strategy, but pollution of air, water, and soil had not entered into their calculations. This had simplified their tasks and avoided immediate costs, although, in the long run, environmental cleanups at some military project sites have proved to be astronomically expensive. CA/T engineers, on the other hand, must avoid violating communities, leveling homes, taking parkland, and polluting with noise and dirty air.

On the basis of an early conceptual design, the Department of Public Works submitted in 1983 the EIS that the Massachusetts secretary of environmental affairs and the U.S. Environment Protection Agency approved in 1985. Subsequently, however, a new and more detailed design responding to environmental-agency suggestions required the preparation of a supplemental EIS into which Joint Venture engineers, environmental specialists, editorial assistants, and employees of the Boston architectural firm of Wallace, Floyd Associates invested months of work.

In 1990, the Massachusetts DPW issued a draft of the *Final Supplemental Environmental Impact Statement/Report* (FSEIS/R).[40] Consisting of several thick volumes, Part 1 describes in more than five hundred pages various environmental impacts. Its seventeen categories suggest the encompassing nature of "environment" as defined by public legislation. They include "transportation," "air quality," "noise and vibration," "energy," "economic characteristics," "visual characteristics," "historic resources," "water quality," "wetlands and waterways," and "vegetation and wildlife."

While some categories are self-evident, others require elaboration. Under the heading of "transportation," the FSEIS/R compares existing traffic conditions with projected ones. Technical characteristics of the venting system for the underground and tunnel sections of the new highways are thoroughly discussed under "air quality." Noise and vibration during construction receive especial attention. Under the heading of "economic characteristics," the FSEIS/R describes the mix of commercial and industrial activity and tourism, as well as employment patterns in the affected areas. It compares existing economic conditions with those expected to prevail after the project's completion. The reduction in traffic congestion, for example, is expected to greatly stimulate retail sales and tourism and to increase employment in downtown Boston.

The FSEIS/R stresses that the project obviates the necessity of any residential relocations; the number of relocated businesses has been held to 134 with 4,100 employees. The project intends to revitalize neighborhoods, especially Boston's North End, which the elevated artery will no longer separate from the downtown center. The architects and landscape architects promise to improve "visual characteristics" of the urban landscape by their design of ventilation and other buildings, lights, and signs, and by landscaping along the project corridor. The project will take special care not to degrade historical structures by vibration and noise. It will survey and explore archaeological sites along the route of the project before construction begins in those areas. The FSEIS/R describes the measures the project will take to minimize harm to parks and recreation areas.

In accord with the law and in order to obtain public participation in the design of the project, the Joint Venture circulated the draft version of the FSEIS/R widely, placing copies in libraries, providing for a public hearing, and defining a public comment period. At a public hearing held 21–22 June 1990, 175 persons spoke, including spokespersons from government agencies such as the U. S. Environmental Protection Agency (EPA) and public-interest groups such as the Sierra Club, and 99 persons provided written commentary. A subsequent period for public comment on the published transcript of the hearing extended from 22 August to 22 September 1990, eliciting 273 letters. The DPW published its response to the various comments.

The heterogeneity of the public-interest groups, whose representatives spoke or provided written comment, suggests the magnitude and complexity of public involvement. The Thousand Friends of Massachusetts, the American Automobile Association, the Archdiocese of Boston/Can-Do Alliance, the Artery Business Committee, the Beacon Hill Civic Association, Bikes Not Bombs, the Boston Building Trades Council, the Boston Society of Architects, the Charles River Watershed Association, the Conservation Law Foundation of New England, and the Haymarket Pushcart Association made their views known.

Speaking for the EPA, Julie Belaga criticized, as did many others, the design of the proposed Charles River Crossing. In company with other civic leaders, Dr. John Silber, president of Boston University, spoke favorably about the reduction of noise and air pollution that should follow the removal of the elevated highway and about the vastly easier access to Logan Airport that would ensue. The Haymarket Pushcart Association expressed concern about the loss of parking spaces and customers during the years of construction; and a representative of the Laborer's Local 223, like other trade union representatives, happily noted the boost to employment the project would provide.

After the formal public review in June 1990 that elicited "scrutiny, questioning, and even second-guessing or downright disagreement," John DeVillars, Massachusetts secretary of environmental affairs, on the basis of the draft FSEIS/R, issued in January 1991 a certificate of approval allowing construction to proceed, assuming that certain mitigation measures would be implemented to reduce negative environmental impacts. He requested that a revised, final FSEIS/R describe in detail the mitigations and other responsive measures that would be taken with regard both to negative environmental impacts during construction and to long-term effects of the project.

Because the Boston Metropolitan District Commission had long planned an urban, riverside park system along the banks of the lower Charles River, an area affected by the bridge and ramps planned for the bridge crossing, and because these would occupy shore and river space, DeVillars called for specific mitigatory measures. These included a suggestion for a year-round-use Olympic-size swimming pool and an ice-skating rink along the river. There had been no swimming in the lower Charles for decades, but DeVillars assumed future cleanups of the polluted stream. He also called for the construction of an aesthetically pleasing bridge structure and for the formation of an advisory panel of prominent architects, environmentalists, and community representatives to participate in the final design of the bridge and associated ramps. In focusing on the Charles River Crossing, he anticipated the emergence of CA/T's major design problem.

DeVillar's certificate of approval also recommended a planning process for the utilization of the twenty-seven new acres in downtown Boston created by the removal of the elevated Central Artery. The load-bearing characteristics of the various sections of the underground artery specified by the engineers would affect the type of building or open spaces at the surface. Besides his concern for parks and new spaces, DeVillars urged the project managers to for-

CONCEPT OF A PORTION OF TWENTY-SEVEN ACRES OF DOWNTOWN BOSTON AFTER REMOVAL OF THE ELEVATED HIGHWAY. *(Courtesy Central Artery/Tunnel Project)*

mulate "creative strategies" for integrating the new highway system with mass transit facilities, for limiting downtown parking spaces, and for reserving highway lanes for high-occupancy vehicles in order to reverse the existing 2:1 auto-to-transit ratio in downtown Boston.

The Massachusetts DPW described its responses in a supplementary volume of the FSEIS/R. Among these, it projected extensive parkland along the banks of the lower Charles River basin to compensate for the necessary presence of the massive bridge and ramp structure. The department also agreed to fund the construction of the Olympic-sized pool and a feasibility study of the potential for ice-skating and public swimming in the lagoons along the Charles. These and other mitigations add considerably to the cost of the Central Artery/Tunnel Project.

Having announced in January 1991 that the revised FSEIS/R complied with the state's Environmental Policy Act, DeVillars nevertheless ordered the CA/T management to continue to consider modifications and options of the proposed design of the Charles River Crossing (Scheme Z). Many environmental organizations and individuals had questioned the design during public hearings. Several months later, the FSEIS/R having fulfilled the requirements of the National Environmental Policy Act, the Federal Highway Administration also signed off on it.

The work of the environmental section of the Joint Venture, however, continued after its completion of the FSEIS/R. The section's staff of about fifty persons, headed by Ed Ionata, had to obtain construction permits from government authorities in cases where the environment would be affected. Ionata and his staff also had to take the lead in preparing another environmental impact statement for a new design for the Charles River Crossing. As unanticipated environmental impacts emerged during construction, Ionata, a graduate of the Yale University School of Forestry who had worked earlier on several major environmental projects, including the cleanup of Boston Harbor, remained in virtual day-to-day contact with representatives of the various local, state, and federal environmental agencies.

]]]]]] **Mitigations**

Before publication of the FSEIS/R, earlier negotiations had resulted in a number of mitigations, or atonements for negative environmental impacts. Salvucci, the Massachusetts DPW, and the Joint Venture engineers negotiated with hundreds of neighborhood, business, and environmental groups, as well as developers and individuals, in response to their putative concerns about anticipated negative environmental impacts. Affluent organizations even hired their own engineers to detail alternative designs for highway alignment, ramps, and locations of ventilation buildings. Less well endowed groups mustered political influence in a manner reminiscent of the successful efforts of neighborhoods almost two decades earlier to block the inroads of highways and to foster rapid transit. Mitigation has been described sympathetically as a consensus-building process; a cynic, however, has called mitigation "blood money." Salvucci in a rare negative moment spoke of responses to demands from Boston City Hall officials as "delivering some chunk of mastodon meat back to the tribe."[41]

Salvucci's position as secretary of transportation and his intimate knowledge of the ethnic politics of Boston neighborhoods aided him immensely in negotiating conflict resolutions and in framing mitigations during the early design phase of CA/T. As head of the East Boston Little City Hall from 1968 through 1970, he had forged an alliance with several socially concerned and politically active Catholic priests. They helped transform the politics of East Boston—where Italians had not been as well organized as the Irish of South Boston—from one of simple job patronage ("Who gave Uncle Louie a job, I'll vote for him") to one of issues, especially the preventing of new highways and Logan Airport expansion from taking homes and causing air and noise pollution. Salvucci had also won lasting respect in East Boston for helping prevent the Logan Airport authorities from routing five hundred trucks during a construction project over East Boston's residential streets: "the kind of street on which kids play, jump rope, and hop scotch."[42] With this strong base of support in East Boston and in his determination to strengthen it, Salvucci, as we have observed, persuaded the Port Authority, which controlled Logan Airport property, to allow the tunnel to surface on its property rather than in East Boston.

Salvucci also took the lead in implementing the construction of the South Boston bypass and seaport access road, and in keeping trucks working for CA/T off neighborhood streets by using barges to haul millions of cubic feet of excavation dirt. Responding to South Boston demands, Salvucci also supported the covering, or decking, of the seaport access road, a major design change that would cost about $20 million. Encouraged by Salvucci, project engineers simplified and reduced the bulk of a complex interchange designed to link the Massachusetts Turnpike, the Southeast Expressway, the seaport access road, and the new tunnel and expressway.

Salvucci obtained the support of Boston's influential Conservation Law Foundation by providing explicit commitments from the state of Massachusetts that the public-transit expenditures declared in the environmental impact statement would be forthcoming in later years under other state administrations. The foundation had also voiced concerns that CA/T would bring more traffic into the downtown area. Salvucci also committed the state to a future freeze on downtown parking. As secretary of transportation, he included these agreements in planning and budget documents in order to commit the gubernatorial administration that was soon to replace Dukakis. Especially concerned that the next administration might not fulfill the agreements with the foundation, Salvucci, the state secretary for environmental affairs, and several other state officials signed an agreement with the foundation in December 1990 that specified expenditures on mass transit and mass-transit station parking as well as restraints on downtown parking.[43] In 1991, the new administration of Governor William Weld accepted the accord with the foundation.

Several years later, CA/T Project officials compiled a database encompassing 1,100 mitigation commitments. These, according to the report, added an estimated $2.8 billion to the total project cost. The commitments include those needed to obtain state or federal government approval, requirements imposed by environmental agencies, and those needed to reduce the impact of construction, such as $450 million for temporary lanes, curbs, and sidewalks to facilitate continued business during construction in the downtown area. (The several-billion-dollar Conservation Law Foundation commitment will not be charged to the project budget.) Other large mitigations singled out by the study include promises of $150 million to the City of Cambridge and $80 million to the Metropolitan District Commission to build the park along the Charles River. Salvucci continues to judge mitigation "a good thing, it's an essential thing."[44]

]]]]]] Scheme Z: Salvucci's Achilles Heel

By late 1990, CA/T appeared poised to move ahead with final design and on to construction. Salvucci was to leave office as the Dukakis administration turned over the reins of government in January 1991, but the project seemed to have the momentum that would carry it forward intact under a new governor.

But at this time public attention focused on the design for the Charles River bridge and ramps. Nicknamed Scheme Z because it was the twenty-sixth alternative explored for the crossing, its engineers and Salvucci hoped it would be the last. Believing it the least objectionable from technical, aesthetic, and environmental points of view, Salvucci had decided in favor of Scheme Z despite the doubts of key members of the CA/T staff and the Massachusetts Highway Department. They announced the plan publicly in August 1988, but the unavailability of three-dimensional models or easily comprehensible drawings of the design muted reactions to it. A year later, the display of a model of the crossing prompted the architectural critic of the *Boston Globe* to label Z "an awful scheme for a great wall across the Charles." An official of the Environmental Protection Agency added that the crossing would be "the single ugliest structure in New England."[45]

Salvucci argued that a rejection of Z would endanger the funding and future of all of CA/T and that the choice of a more expensive alternative, such as one with river tunnels, would lead to the cutoff of funding for the entire project in 1991 when Congress would consider reauthorization of all federal surface transportation programs, including the interstate highways such as CA/T.

Controversy over Scheme Z simmered throughout 1989 and most of 1990. In October 1990, the *Boston Globe* printed an essay by K. Dun Gifford, chairman of a Committee for Regional Transportation, which attacked not only Scheme Z but a number of other aspects of CA/T. Gifford, earlier a confidant of the Kennedy family, chairman of Common Cause, and director of the Conservation Law Foundation, lived in Cambridge and owned a gourmet food business. Memorable metaphors peppered his article. CA/T construction, Gifford wrote, will be like "a jackhammer" opening "Boston's gastrointestinal tract" and will tear out "all arteries, veins, muscles and intestines." "Like any patient in major surgery," he added, "Boston will be comatose throughout the operation."[46]

He envisioned congested surface sections of the artery flanked by high-rise ventilation stacks spewing tons of pollutants into the State House and Beacon Hill. He lamented the failure of planners to provide a rail link between North Station and South Station, predicting that as a result Boston would become a backwater, dead-end city as inland rail routes bypassed it. Scheme Z's "spaghetti bowl" interchange would burden nearby residential neighborhoods with auto exhaust, highway dirt, and road salt.[47]

The chorus of opposition grew louder after Peter Howe, a *Globe* reporter, wrote a December 1990 series on the design of the scheme.[48] Howe learned from several experienced environmentalists and neighborhood activists about a groundswell of protests directed against the design of the crossing. Dan King, a small businessman whose deck on his new house looked out on the proposed site of Scheme Z, was a leader of Citizens for a Liveable Charlestown. He expressed his apprehension about what might happen in his and his neighbors' backyards. Elizabeth Epstein, an active environmentalist and director of the Cambridge Conservation Commission, reported widespread opposition to Scheme Z in her community. Gifford's Committee for Regional Transportation, formed in July 1990 and funded in part by parking lot owner Richard Goldberg, who was protesting about the loss of some of his land to CA/T, entered the fray. He provided substantial legal and public relations support for several protesting groups.

Believing that citizens were not aware of the massiveness of Scheme Z bridges and ramps, the Citizens for a Liveable Charlestown caught the public eye by hiring an artist to prepare an illustration of the "wall of viaducts." Publication in the *Charlestown Patriot* caused a public uproar.[49] Within weeks, other groups, including the Charles River Watershed Association, with more than a thousand members, and the New England chapter of the Sierra Club, joined the chorus.

Reporting that "Red Flags on Scheme Z Rise at City Hall," the *Boston Globe* carried a week-long series by Howe in December.[50] A long article in the Sunday edition on December 9 had a memorable sidebar and map, as well as Howe's attention-catching account of motorists' future problems in using the planned crossing. The *Globe* series demonstrates how effective an alliance of media and activist groups can be in bringing large-scale public works projects under public scrutiny. Howe acknowledges and believes in the power of the press in this role.[51]

Comparing Scheme Z to the "other green monster," the existing elevated Central Artery, Howe projected an image of Scheme Z that covered an area the size of the Boston Common; that consigned strollers and bicyclists passing beneath its structures to the "noise, shadow, and blight of a highway structure as wide as two football fields"; and that created as much new elevated highway as the underground Central Artery was designed to eliminate.[52] Howe did not point out that the "other green monster" scarred downtown Boston and that Scheme Z's bridges and ramps, in contrast, would intrude most directly into an area of railroad tracks and industrial facilities.

Howe asked the reader to see "just how wacky and wild Scheme Z may prove to be" by taking an imaginary drive starting on Route 1 north of the city and heading downtown. Initially, she or he would climb the existing Tobin Bridge to a point almost 150 feet over the Mystic River, then head down into a tunnel 80 feet

THE PRESENT CHARLES RIVER CROSSING (HIGHWAY I-93). *(Courtesy Central Artery/Tunnel Project)*

under Charlestown's City Square, emerge from the tunnel, and climb a ramp with a steep 5 percent grade until a point 110 feet above ground was reached. Then the motorist would cross a five-lane bridge over the river, plummeting 120 feet into a land tunnel in downtown Boston. The imaginary motorist wishing to reach the projected central artery from the heavily traveled Storrow Drive would have to cross the Charles River twice.[53]

By December 14, opponents at a public hearing called Scheme Z a "nightmare," a "travesty," and a "horror show."[54] Noting this rising opposition, the Boston City Council, by unanimous vote, declared its opposition to Scheme Z—but not to the entire Central Artery/Tunnel Project. Salvucci found the reactions overdrawn and the newspaper articles "negative," "unfair," "wordsmith stuff," and their eye-catching graphics distorting.[55] If the public had been shown drawings from a pedestrian's point of view instead of the overhead view used by the *Globe,* Salvucci contends that only a hundred people in the world could probably see the differences in the various Charles River Crossing schemes. Salvucci continued to argue that Z solved traffic and cost problems more effectively than any alternatives, including a design with a Charles River tunnel favored by many opponents of Z. Salvucci explained the need for the massive structures of Scheme Z as a response to the building and parkland constraints of the site, which required concentrating into one quadrant ramps that in the countryside would be distributed over four quadrants (four-leaf clovers). Scheme Z would effect a compromise, or trade-off, solution, Salvucci observed, to the worst highway problem in the country.

Salvucci's arguments for accepting the scheme bring to light the many factors that the designing engineers took into consideration. Scheme Z's many and long ramps responded effectively to federal highway requirements, such as ample weave-sections (sections where motorists are using both on and off ramps within a short distance). During construction, Scheme Z would minimize, as well, detours and interruptions to highway traffic and rapid transit service. The design would produce manageable impacts on potential and actual parkland, leaving 70 percent of the riverbank open (the state had agreed to spend $75 million improving the Charles River Basin, adding new parks, pools, and paths). While Scheme Z, unlike some of the other design proposals, did not provide motorists an easy and direct access to Storrow Drive from Route 1, Salvucci maintained that this would have the beneficial effect of discouraging greatly increased traffic along the drive. If such an increase did occur, this might necessitate the widening of the drive with resultant encroachment on the handsome, bordering riverfront Esplanade.

Scheme Z also avoided river tunneling, which would have stirred up a polluted and toxic river bottom and caused an environmental quagmire. He recalled that the threat posed by Manhattan's Westway highway project to Hudson River aquatic life had killed that project in 1985. An official of the Army Corps of Engineers then observed, "You're not going to see the large project anymore, because we've constructed so many hoops . . . that it's almost impossible to get a large project through." Senator Daniel Patrick Moyni-

MODEL OF THE CHARLES RIVER CROSSING (HIGHWAY I-93), SCHEME Z. *(Courtesy Central Artery/Tunnel Project)*

han from New York said, "The opposition to highways had become so generalized and the techniques for stopping them had become so widespread that . . . we can't get these things done anymore."[56]

On the other hand, Howe expressed surprise that Salvucci, "almost as a matter of personal pride," had tried for several years "to engineer a way to build a critical interchange through a jungle of physical and environmental constraints" without finding any other design more palatable than Scheme Z.[57] Robert Weinberg, a supporter of Salvucci's—and there were many—employed a memorable metaphor to describe Salvucci's tenacious negotiations and trade-offs and the criticism leveled against him during the Scheme Z episode: "The critics are like someone who is watching a guy on a high wire standing on one foot, balancing a grand piano . . . in the air, playing a Beethoven sonata on the piano while sipping soda from a straw and hopping up and down and when he finally gets to the other side, saying, 'He's not a very good piano player, is he?' "[58]

Salvucci was witnessing a sea change in public attitudes toward CA/T: "The project had become Goliath and everybody else was David."[59] In retrospect, Salvucci expresses respect for most of the attacking environmentalists and neighborhood activists, although he believes they usually failed to take into account a sufficient number of technical, environmental, cost, and traffic restraints and that, consequently, they failed to appreciate Z as a compromise solution to a horrendous set of problems.

He reserves his criticism for Goldberg, the part owner of the Park 'N Fly parking lot in East Boston used by patrons of Logan Airport. Goldberg, in Salvucci's opinion, misused environmental regulations, the media, and public opinion to pressure the state and CA/T authorities into responding to his demands for compensation, or mitigation, because a harbor-tunnel interchange required taking some of his property. After Goldberg failed to persuade Salvucci to promise to limit the taking of his property to a ten-foot section, which an engineer hired by Goldberg found feasible, Goldberg employed a well-established Boston law firm and several experienced lobbyists to make his case. He also organized a committee to "Save East Boston"—and his parking lot. Eventually, he and his lawyers evolved the strategy of bringing activist and public pressure on CA/T at its most vulnerable point—Scheme Z. Salvucci sees Goldberg as a prime source of the "firestorm" that broke out around Z in December 1990.[60] But he is at a loss to explain his opponent's underlying motives.[61] Goldberg insisted that Salvucci was "out to get me" because of longstanding East Boston political rivalries.[62]

The editorial pages of the *Globe* continued to support Salvucci and the project, the Howe articles notwithstanding. A heterogeneous coalition that included the 35,000-member Boston Building Trades Council, the Boston Society of Architects, the Construction Industries of Massachusetts, real estate developers, and the Conservation Law Foundation expressed their agreement with Salvucci's call for state and federal approval of the CA/T environmental impact statement and inauguration of the project before Dukakis left office. Advocates of Scheme Z conceded that a review of it for possible design revisions could come later.[63] "If we don't finalize this thing and get into construction, we can lose this project," Salvucci warned.[64]

When approving the FSEIS on January 2, his last day in office, and thereby clearing the way for state and federal permits and licenses to start construction, DeVillars, as the state environmental affairs secretary, at the same time strongly recommended further review of Scheme Z and the exploration of alternatives by a broadly representative design-review committee. Nevertheless, Priscilla Chapman of the Sierra Club of New England rejoined that "this stinks . . . it's full of holes," and Cambridge Mayor Alice Wolf remarked, "I'm not happy. It looks to me that if we want a redesign of scheme Z we'll have to do it through the feds or through court."[65] But DeVillars, justifying his decision, stressed that in downtown Boston a tree-lined boulevard would replace a steel-shrouded strip of macadam and that riverbank parks would grace the new crossing of the Charles River. He became choked with emotion as he publicly defended Salvucci: "No one should lose sight of Fred Salvucci's incredible contributions to serving the public good."[66] Others criticized him for pushing through a flawed design at the last minute.

]]]]]] The New Administration

In January 1991, Richard Taylor, secretary of transportation in Republican Governor Weld's new administration, assuaged various interest groups by establishing the Bridge Design Review Committee. The composition of the forty-two-member committee exemplifies current thinking about participatory design and conflict resolution. John Wofford, the Boston lawyer who had served as the staff coordinator for the transportation study group formed by Governor Sargent in 1969, became the facilitator for the committee. Wofford used open, participatory, multidisciplinary techniques to evolve a consensus.[67] The committee sought to find, through positive compromises, a project design that responded to as many concerns and objectives as possible. Stanley Miller, an area businessman who chaired the committee, regularly reminded the members that tireless effort to reach agreement was in the interest of their grandchildren, who would have to live with the design of this "Gateway to Boston."[68]

The broad, multifaceted, participatory nature of the design deliberations arose from the composition of the committee. Members included representatives from five organizations that were bringing suits against various aspects of CA/T, including Scheme Z. Members also represented national environmental organizations, such as the Sierra Club; local environmental and transportation groups, such as the Charles River Watershed Association, the Committee for Regional Transportation, and the Citizens for a Liveable Charlestown; ad hoc organizations addressing CA/T, such as the Artery Business Committee; and long-standing organizations representing Boston area interests, such as the Boston Chamber of Commerce. Professional engineers, architects, and urban planners representing such organizations as the Conservation Law Foundation, the Boston Society of Landscape Architects, and the Boston Society of Architects also sat on the committee. The Joint Venture made available its multidisciplinary staff, plus five independent consultants including bridge, tunneling, air quality, commuter rail, and traffic experts.[69]

Secretary Taylor had charged the committee with finding the means to reduce the negative environmental and aesthetic impacts of the Charles River Crossing. But instead of revising Scheme Z, in June the committee unanimously voted to abandon it.[70] Salvucci considers the decision a serious error. If the committee and the secretary of transportation had concentrated on modifications instead of the public-pleasing dramatic gesture of abandoning it, then designers, Salvucci believes, could have found ways of eliminating some of the most controversial aspects.

Salvucci recalls that the long-drawn-out committee deliberations "turned it into a nightmare for me watching it."[71] He feels that a modified design for Z could have been introduced without triggering a requirement for a new, time-consuming, and costly environmental impact statement. He acknowledges that in his determination to obtain the certificate of approval from the state environmental office before the Dukakis administration left office, he postponed time-consuming public review of design modifications that might have made the design more acceptable. He passed

on the proposed modification, however, to the incoming administration for processing. He concedes that he may have pushed too
hard for an unmodified Z: "People thought I was nuts, obsessed
with Scheme Z; I had been working too many hours; maybe I was a
little bit nuts."[72]

In its report submitted in August 1991, the Bridge Design
Review Committee proposed a new conceptual design involving a
river tunnel to reduce the number of bridges from three to one and
bridge lanes from sixteen to thirteen, as well as the number of the
looped-ramp structures on the Cambridge side from six to a maximum of four. Committee Chairman Miller stressed that the design
chosen emerged from multidisciplinary discussion involving architectural, land use, and urban design concerning the Charles River
Crossing.[73] The American Institute of Architects gave the new
design an Urban Design Award; several CA/T litigants agreed to
dismiss their lawsuits if the scheme were implemented.

Problems began, however, for the Review Committee's design
when the Massachusetts Highway Department filed a notice of
project change (NPC) and the Massachusetts Executive Office of
Environmental Affairs (EOEA) followed with a decision that the
changes required a new EIS.[74] The design did not move through the
EIS and certification processes expeditiously. The Federal Highway
Administration and the U.S. Army Corps of Engineers called for
consideration and comparison of nontunnel alternatives to the Review Committee's design. Critics warned that digging for a tunnel
would not only be expensive but cause serious pollution problems
for the river. The FHWA cited the traffic hazards and likely high
accident rate in the proposed sharply curving tunnel. The Corps of
Engineers cited its legal responsibility to seek designs that minimized the tunneling of river bottoms polluted with hazardous
waste.

The Joint Venture and several subcontractors then prepared
alternative nontunnel designs and a four-inch-thick environmental
impact statement that compared a revision of the Design Review
Committee's proposal with the alternatives.[75] The EIS focused upon

the aesthetics of the bridge and ramp designs, their impact on land beneath them, and the traffic flow on them. Less massive than Z, the three alternative designs reduced the visual impact of their structures as seen from Cambridge and Charlestown. The new designs also reduced shadows, noise, and air pollution, but at the same time constrained the proposed park and recreation areas beneath them.

Although the EIS found that the alternatives to Scheme Z improved traffic flow on highways interconnected by the bridge and ramps, the statement also discovered that these alternatives had higher predicted accident rates than Z. Scheme Z also enjoyed a cost advantage. The EIS reported few differences when Z was compared with alternatives in the categories of traffic flow, safety, noise and vibration, wetlands, water quality, navigation, fisheries, aquatic resources, threatened and endangered species, historic and archaeological preservation, and hazardous waste. In September, publication of the EIS gave the public and various interest groups opportunity to comment.

James T. Kerasiotes, the new Massachusetts secretary of transportation, promised that he would soon make a choice among the alternatives. The *Boston Globe* predicted that the revised tunnel design with a twelve-lane bridge over the Charles would be chosen. In November, the secretary surprised the public by announcing his selection of a nontunnel, all-bridge design that specified two bridges side by side, one with ten lanes and another with four. The ten-lane cable-stay bridge design, for which Christian Menn, a Swiss bridge architect, acted as chief consultant, calls for cable suspension from two massive 265-foot-high towers. Project engineers acknowledged that "there is no wider traffic structure in the world." An enthusiast proclaimed that "it's one of the most elegant bridge structures in the world."[76] Others characterized the bridge as a signature structure, appropriate as a gateway to a great city.

Kerasiotes stressed in his defense that a timely and sensible decision would impress the federal government upon which the commonwealth continued to depend for 85 percent of the project

MODEL OF THE NEW CHARLES RIVER CROSSING IN THE YEAR 2002. *(Courtesy Central Artery/Tunnel Project)*

MODEL OF THE NEW CHARLES RIVER CROSSING BRIDGE. *(Courtesy Central Artery/Tunnel Project)*

funding: "The federal government is our partner and our banker. We need to convince them of our commitment for integrity."[77]

In response to the Kerasiotes decision, a number of organizations that favored some variation on the Review Committee's tunnel design gave formal notice that they intended to sue to block the secretary's selection. Gifford, head of the Committee for Regional Transportation, who had been a leading critic of Scheme Z, expressed surprise at a decision that flew so in the face of Cambridge and Charlestown critics:

> It's such a silly decision. It ignores the massive citizen participation of the last three years. It ignores the legal realities of the design work of the state highway office itself, which designed a plan far better environmentally than this one.[78]

Alice Wolf, a city councilor and former mayor of Cambridge, observed that "the state has really not kept faith with either the work of the Bridge Design Review Committee or the City of Cambridge. . . . this is going to be here for 100 years. If it takes a year or two longer, so what?"[79] An unnamed project official reacted to the threats to sue: "Cambridge—I don't know what their problem is. . . . They saw what the city of Boston extorted from the project, and they want to get theirs."[80] Seemingly, he or she believed that environmental mitigation shades into extortion.

Kerasiotes maintained despite criticism that "we chose to go over the river because we believe it is more environmentally sound, more sensible." He argued that the tunnel option would take two more years and add $300 million more to the cost of CA/T. Already the two-year search for an alternative to Scheme Z had delayed and added millions of dollars to the cost of the project; now Kerasiotes estimated that the design he had chosen, even though cheaper than the others, would drive the cost of the project up $1.3 billion to a total of $7.7 billion. The completion date for the project moved to 2004. The Environmental Protection Agency and the Federal Highway Administration approved the nontunnel alternative in 1994; the Joint Venture proceeded with detailed design.

ARTIST'S CONCEPT OF THE NEW CHARLES RIVER CROSSING, SOFTLY ILLUMINATED TO BLEND WITH THE LIGHTING OF THE BOSTON SKYLINE. THE CA/T PROJECT HAS INVESTED IN LANDSCAPE AND URBANSCAPE AESTHETICS. *(Courtesy Central Artery/Tunnel Project)*

]]]]]] Management

Turning to the Joint Venture's management structure, we focus on 1993, when the project moved from a period of ill-defined, unpredictable growth to one when the scope of the project became contained and predictable. Design shifted from the preliminary to the final phase and construction began on the harbor tunnel.

In 1993, two Bechtel employees acted as top managers for the Central Artery/Tunnel Project. Theodore G. "Tad" Weigle, the program manager, joined the Joint Venture in 1992 after having come to Bechtel as a vice president in 1990 from Chicago. There, as executive director of the Northeastern Illinois Regional Transportation Authority, he had been responsible for all public transportation matters in Chicago. For the preceding five years, having served as the chief operations officer of the Washington, D.C., metropolitan transportation system, he was in charge of the subway and bus networks. Weigle's familiarity with legislative constraints, financing, and the management of operating systems complements the construction management experience of Jerry Riggsbee, the project manager. After seven years as an independent construction contractor, for the next eighteen years he managed the construction for Bechtel of canals in California and airports in the Near East and elsewhere.

Weigle has a college degree in education and studied at the Harvard Business School, while Riggsbee obtained a degree in civil engineering from the University of California–Berkeley. Weigle concentrated on the project's external financial and political problems, while Riggsbee focused on engineering problem solving within the organization. Together they constituted what they call a "one-stop shopping center for general management responsibilities."[81]

They exchanged views regularly with their counterparts from the Massachusetts Highway Department and the Federal Highway Administration during regular Tuesday "critical issues" meetings held at CA/T headquarters at One South Station in Boston. Project core managers and engineers also took part in the discussions of critical problems that arose during design and construction. These included failures to meet schedule milestones and cost overruns. These managers and engineers met earlier to prepare the agenda of critical problems for the following Tuesday. Schriever had similar meetings during the Atlas Project, though his fell on "black Saturdays."

The Joint Venture managers look for oversight to their client, the Massachusetts Highway Department, whose chief representative is Peter Zuk, the project director. Trained in law and experienced in the legal intricacies of construction contracting, Zuk was brought to CA/T from the State Attorney General's Office in 1991. His primary responsibility is to oversee the work of the Joint Venture, to keep close tabs on schedule and costs.[82] Having control of funds, Zuk considers appeals from the Joint Venture because of design changes or mitigation demands for funds beyond those specified by contract. Even though he and his small group of Massachusetts Highway Department employees are formally clients, he and they have over time worked so closely with Joint Venture management that they function like partners presiding over the project.

PETER ZUK, CA/T PROJECT DIRECTOR. *(Courtesy Central Artery/Tunnel Project)*

Zuk takes responsibility for negotiating with Boston and state agencies who make various demands upon the project. Given his acquaintance with technical matters and his political sensibilities, he is particularly effective in dealing with organizations and persons who see the Central Artery/Tunnel Project as a "cash cow."[83] Since 1992, when the scope of the project became relatively well defined, Zuk and his deputy project director have been committed to cost containment.

Zuk depends on his deputy project director, William S. Flynn, a retired Army lieutenant general, to maintain the project on schedule and to contain costs. Flynn came to the project in 1992. Having risen in the Army from an infantry platoon leader, he, like General Schriever a product of the ROTC rather than West Point, rose to head the Army's Tank and Automotive Command, whose 6,000 engineers and technicians had responsibility for developing new vehicles such as the M1 tank. Flynn believes that a person heading an engineering organization needs first and foremost to know how to lead people toward project goals. To do this, according to him, an engineering degree is not essential; leadership qualities such as one can learn from Army command are. From Flynn's perspective, the system builders whom we have encountered, such as Schriever, Ramo, and Salvucci, are leaders, not simply managers. He also finds that providing leadership for the engineers who are "digging a hole" differs little from performing this role for those making a tank. Asked if he could have responded to the political problems that Schriever found so frustrating when he turned his attention to designing civil systems, Flynn quickly rejoins that he leaves "politics" to Zuk, who has proven himself a master of messy complexity.[84]

In his efforts to hold costs and maintain schedule, Flynn says he has to "beat hell out of the Joint Venture engineers" while at the same time "loving" them for their impressive accomplishments. Well trained and with high professional standards, the engineers want to build a gold-plated "Mercedes," which Flynn understands and appreciates, but he constrains and disciplines them so that they will build a solid, affordable economy car. He and Zuk realize that otherwise CA/T may never be built as conceived. Flynn helps hold the 1,000 or so Joint Venture engineers and technicians to schedule and costs by insisting that as individuals they define and accept well-defined responsibilities.

]]]]]] Design Teams

To manage a project so complex as CA/T, managers break down the project into conceptual design, preliminary design, final design, and construction phases. Conceptual design lasted from 1985 to 1990, most of the preliminary design was done from 1990 to 1993, and final design, overlapping with preliminary design, lasted from 1990 into 1997. Conceptual design provided enough of an overview of tunnel and Central Artery alignment for the preparation of an environmental impact statement and for general planning.

Because the Commonwealth of Massachusetts does not permit the same firm to do design and construction, the Joint Venture does the preliminary design and manages contractors responsible for final design and construction. In 1993, the project had 56 final-design sections, or packages, with a prime contractor for each. At the same time, the project had 132 construction packages for which contracts were, or were to be, issued. Many of the design packages correspond to construction packages—a design package and a corresponding construction package for the South Boston bypass road, for instance. For other sections of CA/T, there is a single design package but several construction packages. This is the case for the Logan Airport interchange. Prime contractors share their responsibilities with subcontractors and consultants. We witnessed the same approach to complexity taken by the managers of the Atlas Project, who also divided a system into subsystems. As a result, the managers of CA/T face similarly pressing problems of coordinating the design and development of subsystems, or packages.

Jeff Brunetti, head of the preliminary design team in 1993, has been with Bechtel for sixteen years and on the CA/T project for four. After graduating in structural engineering from the University of Colorado, he helped design airports and other major structures abroad, becoming a construction manager for Bechtel in Saudi Arabia. He is one of a new breed of engineers who feels comfortable with a project greatly shaped by public participation and environmental law. Some of the older engineers, accustomed to concentration on technical and economic factors, find a CA/T learning curve tough, even forbidding.[85] Brunetti, however, although he finds CA/T unprecedentedly complex, finds the new dimensions of engineering "frustrating at times, but lots of fun."

JEFF BRUNETTI (RIGHT), HEAD OF THE PRELIMINARY-DESIGN TEAM FOR THE CHARLES RIVER CROSSING. *(Courtesy Central Artery/Tunnel Project)*

Brunetti, his Joint Venture team, and a number of outside consultants, including specialists in architecture and landscaping, several hundred people altogether, do background studies and prepare the preliminary design before the final design is turned over to outside contractors. Packages amount to between $200 million and $300 million each. When the preliminary designs are turned over to the final designers, design is 25 percent complete.

In order to expedite the process, potential final design firms, all of whom must by law have local offices, prepare competitive proposals before the preliminary design is complete. So that these firms can prepare their proposals effectively, Brunetti's team turns over to the would-be design contractors between two hundred and four hundred sheets of drawings plus hundreds of pages of detailed descriptions of electrical, mechanical, and other systems. His team also releases to them hundreds of pages detailing environmental commitments. With the advice of the Joint Venture, the Massachusetts Highway Department selects the winning bids.

TONY LANCELLOTTI, CA/T DEPUTY PROJECT HEAD. *(Courtesy Central Artery/Tunnel Project)*

]]]]]] Final Design

As the project moved from the preliminary to the final-design phase, the problem of managing final-design contractors became paramount. Earlier, Brunetti directed the preliminary design team and Tony Lancellotti had charge of final design; in 1993, Brunetti became preliminary and final-design manager for the Charles River Crossing and another critical section of the project, and Lancellotti became the deputy head of the entire project. Lancellotti received his undergraduate and graduate degrees in engineering at Columbia University, then joined Parsons Brinckerhoff. Before coming to the Joint Venture in 1987, he worked for six years on the Fort McHenry Tunnel in Baltimore.

During the intense final-design phase, he and his staff met with representatives from the state environmental protection agency often to make certain that the design specification accorded with environmental regulations. If design had violated these, then the government could have called construction to a halt and imposed various penalties on the project.[86]

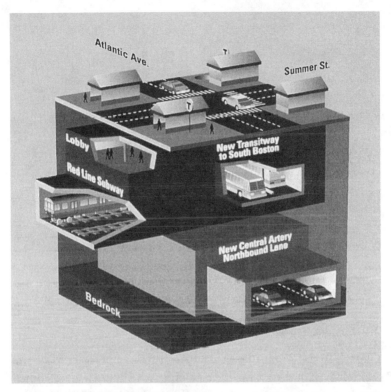

DRAWING OF THE UNDERGROUND ARTERY PASSING BENEATH AND ABOVE RAPID-TRANSIT LINES AT
DEWEY SQUARE. *(Courtesy Central Artery/Tunnel Project)*

Managing the final-design contractors responsible for the various packages, or sections, of artery and tunnel requires close coordination. The contiguous sections must dovetail nicely at interfaces. Lancellotti and his team needed also to resolve disagreements that arose among final designers at these interfaces. The Joint Venture helps the contractors locate sources of information when problems arise outside the contractor's area of competence. Ultimately, the final-design contractors produce plans, specifications, tens of thousands of drawings, and cost estimates needed by firms bidding for construction contracts. Final design and construction proceeded simultaneously until 1997, when the design was mostly completed.

A SECTION OF THE FUTURE UNDERGROUND ARTERY BENEATH THE DOWNTOWN AREA, CLEARED OF THE ELEVATED ARTERY, IN THE YEAR 2004. *(Courtesy Central Artery/Tunnel Project)*

]]]]]] Coordination and Scheduling

Like the Atlas Project, the Central Artery/Tunnel Project has a control center to deal with the critical job of scheduling and coordinating the activities of the various sections of the Joint Venture and of the numerous contractors and consultants. Information from them flows into a control center where computers process it to provide visual representations of interface coordination, critical development paths, and progress milestones. CA/T engineers, like most of their professional brethren, feel more comfortable with visual representations than with verbal reports.[87]

A staff of about thirty persons, mostly engineers, manages the coordination and scheduling activities that take place within the CA/T control room. The staff presents charts of processed information that ranges from detailed information to general overviews. They prepare the general information charts pertaining to design and construction for senior managers and engineers. Along the vertical of these charts the staff lists the various design and construction packages; the horizontal line denotes time—1991 to 2004, for example. So, a bar extending out horizontally from one of the design packages—"Logan Airport Interchange," for instance—conveys the state of the project by the degree of extension along the time bar. The staff provides the information on the more detailed charts mostly for the design and construction contractors, so that they can have a clear picture of their progress and problems, and be able to adjust their work plans regularly.

As in the Atlas control center, those in charge mark progress milestones on their charts. When reports from contractors reveal that they will not reach a milestone on time, the control center alerts Joint Venture engineers and managers of the emergence of a critical problem. If a contractor fails to reach a milestone as scheduled, then other contractors who depend on completion of prior work cannot proceed with theirs and the entire project may be thrown into arrears. As we noted when discussing the Atlas project, failure to reach a scheduled milestone in one package or section of the CA/T system creates in the advancing front a reverse salient which, because of the interdependency of the sections, must be corrected in order for coordinated progress to continue.

The CA/T control center also tracks income and expenditures so that shortfalls can be anticipated. In this connection, the center takes responsibility for monitoring congressional and state legislative decisions that will affect the flow of funds to the project. The Highway Department has to approve expenditures that will exceed the budget. These and other decisions related to coordination and scheduling are called "critical path management."

]]]]]] Design of the Advanced Highway Management System

For what is described as "the largest, most complex underground highway system in the world," CA/T management is designing an advanced highway management system.[88] Civil engineers predominate in most Joint Venture departments; an exception is the Highway and Control Systems group, dedicated to the design and deployment of the traffic management system for CA/T. The design competence of the electrical or communications engineers in this section is not limited to control systems for highways, but includes power, chemical, and manufacturing plants and installations for airlines, railways, and rapid-transit systems.[89]

In the case of CA/T, the need to coordinate traffic surveillance, accidents and breakdown detection, and rescue and removal facilities stimulated the group to develop centralized control. A typical tunnel emergency may involve a vehicle fire, inoperative vent fans or fans spreading the fumes, or the failure of the regular lighting system. The operators in the central control room need sensors, or detectors, to locate the accident and to monitor the vent fans and lighting. They also require controls to change the venting pattern and to turn on emergency lighting. Prompt location of accidents requires constant monitoring of the rate of traffic flow in all lanes in order to identify slowdowns that indicate accidents. Various sensing devices such as radar and television cameras, located in tunnels and elsewhere, also help locate accidents. In addition, the control room must call and direct the rescue vehicles, meantime directing traffic flows to provide access to the disabled vehicle.

Sensors, communication networks, and computers are critical control-center components. Over 15,000 sensors and control points and 350 TV cameras are included in the design for CA/T. Electrical sensors (loop detectors) embedded in the highway will provide the data needed for the analysis of traffic flows involving the 250,000 vehicles expected to use CA/T daily. Located at regular intervals along the highway, they send signals to the control center as vehicles pass over them. From these data sent to the control center by fiber-optic cable, computers extrapolate traffic flows. Computers also provide operators with a menu of possible responses to emergencies.

]]]]]] Wallace, Floyd Associates

In addition to engineering design and management, CA/T involves architectural design, for which the Boston architectural firm of Wallace, Floyd Associates Inc. has responsibility as a subcontractor to the Joint Venture. The architects are fond of pointing out that

VIEW OF DOWNTOWN BOSTON OBSTRUCTED BY THE ELEVATED ARTERY. *(Courtesy Central Artery/Tunnel Project)*

ARTIST'S CONCEPT OF THE SAME VIEW WITH THE ARTERY UNDERGROUND. *(Courtesy Central Artery/Tunnel Project)*

the Joint Venture engineers design what the public will not see, but that they are responsible for what the public will see. The firm has the "lead responsibility" for architecture, landscape architecture, and urban design. Further, it takes part in the preparation of environmental impact statements, sharing the responsibility for maintaining positive community relations and helping to mitigate negative environmental impacts.

Peter Floyd and Dave Wallace established their firm in the 1960s, Floyd having been an architectural assistant to Buckminster Fuller, the famed inventor, designer, and philosopher-guru. The firm designed a number of Fuller-type geodesic domes for the Defense Department and, with Fuller, the geodesic dome at the Montreal World's Fair. When the Massachusetts Department of Public Works called for design and construction bids on the CA/T project, the firm joined with Bechtel and Parsons Brinckerhoff to make a proposal with the Joint Venture as prime contractor, and Wallace, Floyd as the architectural subcontractor. By combining large engineering firms of international reputation with a small local architectural firm intimately familiar with local circumstances, the proposal responded to a complex set of needs.[90]

Like Salvucci, the principals at Wallace, Floyd knew that the Irish and Italians responded vociferously and collectively to threats to their homes and neighborhoods. Sensitivity to the aesthetic and environmental concerns of Cambridge elites and to the economic interests of downtown Boston were not to be discounted either. Wallace, Floyd could help adjust the macroperspective of the Joint Venture to the microconcerns of Boston neighborhoods.

Representing Wallace, Floyd, Hubert Murray became chief architect on the project. Murray and his team did preliminary architectural designs and also monitored the final design for a number of CA/T buildings, including six large buildings housing equipment for ventilating the harbor tunnel and underground sections of the highway. The team also had architectural, in contrast to engineering, design and construction responsibility for viaducts (bridges and elevated ramps), tunnel interiors and portals, and pedestrian walks. The Wallace, Floyd team customarily worked side by side with Joint Venture engineers at the project headquarters at One South Station in Boston.

The technical concerns of Joint Venture engineers, various interests expressed in the public review process, and federal highway regulations constrained the architectural choices of Wallace, Floyd and its associates. The trade-offs and negotiations required of architects involved in a large public works project like CA/T distinguish them from the heroic and mythical architectural type portrayed by Ayn Rand in her novel *The Fountainhead*.

After the start of the project, Wallace, Floyd turned over part of its architectural responsibilities to other Boston firms: some to Stull & Lee, and some of the landscape architecture responsibilities to Carol R. Johnson & Associates. Both firms qualified as disadvantaged business enterprises (DBEs), Stull & Lee being headed by an African American and Carol Johnson by a woman. To qualify as a DBE, a firm has to be at least 51 percent owned and controlled "by one or more socially and economically disadvantaged individuals." These include U.S. citizens who are women, black Americans, Native Americans, Hispanic Americans, Asian-Pacific Americans, "or any other filling the criteria."[91] In addition, the firm must have an annual turnover below a standard dollar amount. By federal law, DBEs must be awarded a substantial number of contracts for federally funded projects, or "have the maximum opportunity to participate" in U.S. Department of Transportation–funded projects.[92] Wallace, Floyd relinquished about 50 percent of its contractual assignments to the DBEs, but took responsibility for monitoring and coordinating their work. According to Murray, the troika functions well.[93]

Designing "what the public would see" includes, in addition to buildings, the signage, light fixtures, and tunnel tiles. During the design process, the Murray team explored the alternatives of giving the motorist on the Central Artery the impression that she or he is driving either on one separate highway or on the junction of two—Interstate 93 and Interstate 90 (Massachusetts Turnpike).

The team opted for some unifying features such as signage and light fixtures but left open the possibility of using color to distinguish between the I-93 and I-90 sections of CA/T. The firm must decide whether the various structures of the system should look the same in East Boston, South Boston, and downtown Boston, or should take on the character of the neighborhoods and districts through which it passes.

In the case of the ventilation buildings, Wallace, Floyd designed their exteriors to suit the local surroundings: brick and stone for downtown near the historic Bulfinch Triangle; masonry and metal paneling to give an industrial look near the airport. Because some of the design choices for signage were not in accord with standard federal highway practice, the design team had to win over the FHWA.[94]

In designing the exteriors of the ventilation buildings, the architects responded to a host of design constraints. The design of ventilation building number four, in the heart of downtown Boston close to the Government Center, acknowledges local physical constraints, environmental requirements, public review pressures, the influence of various political and business interests, and nationwide standards imposed by the Federal Highway Administration.[95] Because the building would eliminate an existing 250-car parking lot, Boston North End residents and businesses insisted on new, compensatory underground parking beneath the building. Because the construction of the building would displace the small street establishments vending fruits and other commodities, the project promised the "Haymarket Traders" space near the building. The building had also to house a subway entrance as well as have the structural support that allowed the building to be raised above the tracks. In addition, the public wanted a pedestrian walkway through the building, and the Boston Redevelopment Authority insisted on commercial space to bring people to the area. The designers also promised that the ventilation building would preserve the visual easement which allowed Bostonians a view of the Old North Church from the neighboring subway station.

]]]]]] **Management Review at the Point of No Return**

With the project moving into the construction phase, Zuk deemed it wise to have a thorough evaluation of the management of the project by Peterson Consulting of Boston; Lemley & Associates of Idaho, a consulting firm deeply involved with the management of the English Channel tunnel; and Professor Jay Lorsch of the Harvard Business School. The review took place during several months in 1994 and 1995, with a final report submitted in September 1995. The report had two parts, one on the Joint Venture with its 1,000 employees and another on the Massachusetts Highway Department CA/T team of about 50 persons. The review team interviewed more than 100 Joint Venture and Highway Department persons and scrutinized, as well, countless documents.

When the report became public in December, *Boston Globe* headlines read "Mismanagement Blamed for Costs, Delays in Big Dig."[96] Relying on unidentified sources as well as the report, the reporter, Thomas C. Palmer, Jr., who usually wrote for the newspaper about the project, referred to Joint Venture managers allowing "costs to spiral needlessly," and the project being "on-the-job training" for the inexperienced Highway Department staff. A week later, Palmer quoted "sources" who described "a top-heavy management partnership" and a Massachusetts Highway Department team that was "largely inexperienced in such complicated work."[97] Zuk was quoted as saying that such comments are "far out of proportion. . . . There are few $8 billion projects. . . . There are fewer $8 billion projects in a downtown urban environment." (Despite the negative comments, Zuk and Flynn are characterized in the *Globe* as generally skilled leaders.)

The report itself spoke more qualifiedly. It identified "areas of ineffective management, inadequate controls, and a deteriorating relationship between the Joint Venture and the Department," but went on to observe that the Joint Venture and the Department "have [already] implemented the majority of the recommendations stated in this report."[98] (The recommendations had become known to the Joint Venture and the Highway Department months before publication of the report.)

The reviewing team's major recommendation called for the abandonment of the "functional organizational structure" and its replacement by "an area-responsibility organization." Functional organization subdivided management into functional departments such as engineering (design), procurement, project services, and construction. Beneath each manager of a functional department were area managers who presided over preliminary design, final design, and construction contracts in the various geographical sectors of the project. Concentrating overall responsibility for the various areas in the functional managers resulted in fragmentation of control, in the opinion of the reviewing team.

The area-responsibility concept recommended by the team places overall responsibility on about a half-dozen geographical area managers rather than on functional managers. Area managers report to the project executive, who heads the Joint Venture. They have responsibility for coordinating design, construction, and other functions in their respective areas, such as East Boston and South Boston, as well as for maintaining schedule and controlling costs. (Under the old arrangement, one functional manager had prime responsibility for design in all areas and another for construction in all areas.) Reporting to each of the newly created area managers are several project managers, who have responsibility for project sectors, or contract packages, within the geographical area. Generally, packages correspond to geographical sectors and facilities of the project; for example, the South Boston bypass road and the Logan Airport interchange. We should recall, however, that the responsibility for the Joint Venture is managing, or supervising, contractors who do the actual final design and construction. In effect, the review team was striving for integration and coordination of functions by making area managers task oriented, or mission oriented. Zuk characterized this concentration of responsibility as single-point responsibility. "Someone should go home at night," he adds, "with a stomachache about making sure something occurs. Senior management always has stomachaches, but must parcel out some of these to other people."[99]

Besides emphasizing the need for the area-responsibility approach, the review team stressed the need for the Joint Venture to structure its management policies around the project schedule. Earlier, we noted the existence of a control center where engineers tracked the progress of the project and kept tab on costs. The management review team recommended that CA/T management focus its attention on the schedule generated by computer programs and graphically displayed by the control center, and that it use the schedule as the "primary tool to provide a common thread throughout the project."[100] The schedule reveals the interrelationship among the overall project areas and sectors; it also identifies reverse salients in the expanding front of the project—those parts of the project that are not on schedule.

In addition to pressing for the maintenance of schedule, the review team stressed the need to contain costs, relating them to mileposts along the schedule. More specifically, when a task, such as the construction of a particular tunnel segment, or package, is completed, the management should know its precise cost. We have observed that Zuk and his deputy Flynn supported a management approach requiring that the Joint Venture engineers not exceed specified costs for a particular task and that they complete the task within a specified time.

The commonwealth showed its determination to contain costs when in November 1994 it insisted upon the replacement of Weigle, the head Joint Venture manager, after he publicly appeared to contradict Transportation Secretary Kerasiotes's statement of a month before that the cost of the CA/T Project would be firmly held to $7.7 billion plus inflation. Weigle told the *Boston Globe* that uncertainties in downtown construction might push costs up. Zuk said that Weigle's transfer came because the project was entering a new phase, with emphasis on an intense period of downtown construction, so a "hands-on construction manager" was needed. Other state officials contended that the project under Weigle had not been sufficiently committed to holding down costs.[101]

In summary, the review team's report struck a positive note by characterizing the Joint Venture personnel as experienced and competent and the project as part of "a significant trend of transferring management of publicly funded projects to private enterprises," a trend that has become "increasingly popular for megaprojects, because it allows state and federal agencies to bolster their resident technical and professional resources with the highly specialized expertise from the private sector."[102] In the epilogue, we, too, shall portray the Central Artery/Tunnel Project as manifesting a future trend—toward an open postmodern style of coping with complexity.

VI Networking:

ARPANET

Computing may someday be organized as a public utility just as the telephone system is a public utility. . . . The computer utility could become the basis of a new and important industry.

John McCarthy (1961)

▼ The nationwide, real-time, interactive, computer-based information network, the ARPANET, became the first of an increasing number of information networks that spread across the United States and beyond after 1969. Soon interconnected, these networks formed the Internet, to which millions of computer users are connected today. Funded by the Advanced Research Projects Agency (ARPA), a U.S. Defense Department agency, and developed by university research centers, the ARPANET research and development project began in 1969. It culminated in 1972 with a public demonstration of a small network that interconnected computers, primarily at university sites. The last of the technological systems whose creation we shall consider suggests the characteristics of future project management and engineering.

The history of ARPANET provides a memorable, salient example of the manner in which ARPA, using a light touch, funded and managed the rapid development of high-risk, high-payoff computer projects, especially in the 1960s and 1970s. The ARPANET project also provides an outstanding example of federal government funding of academic scientists and engineers intent upon nurturing a new field of knowledge and practice, in this case a computer network. While the presumed military threat from the Soviet Union during the Cold War seemed to justify the military-funded SAGE and Atlas projects, the justification for military funding of the ARPANET project is not so obvious. Military needs can be discerned in the background of the ARPANET picture; the foreground contains scientists and engineers motivated by the excitement of problem solving and the satisfaction of advancing a burgeoning field of computer communications.

The academics who were developing the net were often directing university computer research centers populated not only by fellow faculty members but also by aspiring graduate students; the progenitors of ARPANET contributed substantially to the buildup of computer science departments at major research universities. ARPA funding created a body of knowledge, a set of techniques, and, as Professor Leonard Kleinrock, one of the principal developers of the network, adds, "a cadre of talent" that proved to be "very significant for the United States."[1]

Through the history of ARPANET, we shall discover government funding playing a critical role in one of the opening acts of the so-called computer and information age. The part played by the government, namely the military, in this historical technological transformation raises a perplexing question: is government funding needed to maintain the revolutionary development of computing and is government funding needed to generate other technological revolutions in the future?

In following the history of Atlas, we learned much about the way in which a government agency, the Western Development Division of the Air Force (WDD), acted not only as funder but also as manager of a research and development project. In both the Atlas and the ARPANET projects, a diverse set of nongovernment organizations acting as contractors carried out the bulk of the research and development. A prime function of WDD and ARPA was to schedule and coordinate the activities of the heterogeneous set of contractors. In both cases, the government as project manager granted the contractors considerable freedom to fulfill their responsibilities as long they met the specifications required for their particular component of the overarching system. This avoidance of micromanagement allows us to speak of a light touch on the part of the government project managers. WDD, as we have seen, and ARPA, as we will see, assumed hands-on roles only when schedules and specifications were not being met.

The major difference between the Atlas and the ARPANET projects was the much larger size of Atlas and this project's use of a systems engineering organization, Ramo-Wooldridge Corporation. Atlas involved seventeen principal contractors and hundreds of subcontractors, while ARPANET had less than half the number of principal contractors and a handful of subcontractors. These differences in size resulted in some contrasting managerial techniques. ARPANET's informal organizational structure and the spontaneity of its problem-solving approach suggest the engineering and management style of small groups working within the overarching structure of a large project like Atlas.[2] Another difference is that ARPANET was initially built on top of another major system, the Bell telephone system, which provided the communications network.[3]

]]]]]] Military Interest in Command and Control

The interactions and evolution of events, institutions, and people that culminated in the deployment of the ARPANET can be traced back to the American military's interest in fostering the development of command and control techniques, sometimes called command, control, and communication (C3). We encountered command and control in our history of the SAGE air defense systems and in the introduction of the Whirlwind computer into this system. We also noted that the MITRE Corporation developed a number of command and control systems. The Whirlwind made possible real-time information processing, or the display of information about the movements of aircraft as these occurred. With the Whirlwind, people could interact with the information being processed and displayed on the computer screen; operators could select and highlight desired real-time information from amid the mass of other information being fed into the computer from radar and other sensors. Operators could also interact with the computer memory by asking for the display of information stored there. Military commanders used the information processed and displayed to make decisions and give commands intended to control the air defense situation.

As Air Force interest in computer-dependent command and control increased in the late 1950s, the Defense Department turned to ARPA for the cultivation of computer development. (Later the name changed to DARPA, Defense Advanced Research Projects Agency, but ARPA will be used throughout this chapter.) The Defense Department established ARPA in 1958 in response to the post-Sputnik concern in the United States that our military research and development in particular and our engineering and science in general were falling behind that in the Soviet Union. This organization had the responsibility of funding cutting-edge research and development. ARPA programs focused on problems associated, for instance, with ballistic missiles (reentry)—a particularly intractable problem, as we have seen—on ballistic missile defense, and on nuclear test detection. By the late 1960s, computer-related projects funded by ARPA, including ARPANET, cost in the vicinity of $30 million annually.

In 1959, in response to the military's interest in command and control, ARPA awarded a $6 million contract to the System Development Corporation (SDC) to do conceptual and operational studies related to command and control, including the interaction of the operators and the computers. SDC had developed complex software for SAGE computers, and, in so doing, had trained hundreds of programmers and helped launch the computer-programming community in the United States.

]]]]]] J.C.R. Licklider

Joseph Carl Robnett Licklider transformed the ARPA command and control program into an encompassing endeavor that greatly stimulated the development of computers and information systems not only for the military but for academia and industry as well. The

JOSEPH CARL ROBNETT LICKLIDER AT SYMPOSIUM ON ENGINEERING APPLICATIONS OF SPEECH
ANALYSIS AND SYNTHESIS, MIT, 1953. *(Courtesy MIT Museum)*

first head of the Information Processing Techniques Office (IPTO)
of ARPA from 1962 to 1964, Licklider had a powerful vision of
interactive, time-sharing, and networked computers. With a Ph.D.
in psychology from the University of Rochester in 1942, he did
research in the 1940s in a psychoacoustic laboratory and taught as
a lecturer in the psychology department at Harvard University.
During World War II, the psychoacoustic laboratory studied the
maintenance of vital human communications under the noisy and
mobile conditions of combat.[4] After the war, Licklider moved to
MIT to head an acoustics laboratory there and to fill an appointment as associate professor in the electrical engineering department. He also served as a research associate in the MIT Research
Laboratory of Electronics, noted for its espousal of interdisciplinary
research. Licklider gained a full exposure to the MIT style of
research in the 1950s, one intensified by his participation in Project
Charles, the summer study group that helped define the SAGE
Project. He said that he "fell in love with the [MIT] summer study

process."[5] He also enormously enjoyed being part of the circle of scientists and engineers who met regularly in a discussion group presided over by Norbert Wiener. Licklider's MIT experiences were rounded off by his appointment as a group leader at the MIT Lincoln Laboratory.

In 1957, he moved to Bolt Beranek and Newman (BBN), a high-technology firm founded in 1948, to establish a psychoacoustics laboratory there. MIT professors Richard Bolt, Leo L. Beranek, and Robert B. Newman had founded BBN initially as a part-time venture to do architectural acoustics. Staffed by a preponderance of Harvard and MIT scientists and engineers, the firm specialized in solving acoustics problems amenable to statistical analysis techniques that involved computers. The move from acoustics through statistics to computers helps explain BBN's leading role in the computer field by the late 1960s. Heading the department of psychoacoustics, engineering psychology, and information systems research at BBN, Licklider focused his research on computers by investigating the interaction, the diagrammatic representations, and the simulation of complex systems, and by supporting the development of computer time sharing by BBN. Time sharing, which provided access for a number of users to a single mainframe computer, paved the way for the networking of a number of computers.

Licklider's success in furthering the development of computer and information systems, however, lies not so much in his cultivation of research and development as in his role as a visionary who inspired the problem choices and research and development activities of numerous computer and information pioneers. These, in turn, became members of a network that he helped establish and cultivate. His imaginative reasoning and programmatic thinking are impressively laid out in a seminal paper, "Man-Computer Symbiosis," published in 1960. This paper is a clarion call urging that computers be thought of as means of enhancing human thought and communication, not simply as arithmetic, or calculation, engines as they mostly were at the time.[6] In this essay, he calls for a partnership—"very close coupling"—between humans and electronic computers that would facilitate thinking, decision making, and the control of complex situations. He does not dispute the claim of his colleagues in artificial intelligence that in the distant future computers might dominate cerebration, but he thinks that in the near future, perhaps the next ten to twenty years, a symbiotic partnership would evolve instead. These years, he anticipates, "should be intellectually the most creative and exciting in the history of mankind."[7] Such a vivid pronouncement helps us sense the enthusiasm he generated among the converted.

Licklider opened his "Man-Computer Symbiosis" with this analogy:

> The fig tree is pollinated only by the insect Blastophaga grosso-rum. The larva of the insect lives in the ovary of the fig tree, and there it gets its food. The tree and the insect are thus heavily interdependent: the tree cannot reproduce without the insect; the insect cannot eat without the tree; together, they constitute not only a viable but a productive and thriving partnership. This cooperative "living together in . . . intimate association or even close union of two dissimilar organisms" is called symbiosis.[8]

Man-computer symbiosis, he adds, is a subclass of man-machine systems. Other human-machine systems use machines as extensions of humans. Still others deploy humans to extend machines, to perform functions, for instance, that cannot yet be automated. By contrast, man-computer symbiosis depends on an interactive partnership of man and machine.

He describes how a partnership would change the way computers would be used, but first he points out the laborious, time-consuming way in which they were then being used. To solve a problem using a computer, the researcher had first to formulate the problem; then he or she had to turn to a professional programmer to program the problem for the computer; after this, the problem written in computer language was submitted to the operators of a centrally housed computer who placed the program in a queue to be run as computer time became available; and, finally, the computer processed the information, then printed out the results. This procedure was known as the "batch process."

Instead of this process, Licklider believed that problems in the future "would be easier to solve, and they could be solved faster, through an intuitively guided trial-and-error procedure in which the computer cooperated, turning up flaws in the reasoning or revealing unexpected turns in the solution."[9] Many problems, he adds, could not only be solved, but also formulated, with the aid of a computer.

Despite his call for a symbiotic relationship, Licklider, unlike contemporary artificial intelligence enthusiasts, relegated the computer to tasks mostly clerical or mechanical. He had found by self-observation that he gave most of his so-called thinking time to clerical or mechanical activities involving searching, calculating, plotting, transforming, and determining the logical consequences of hypotheses, and otherwise preparing the ground for his occasional—but essential—insights and decisions. All but the insights and decisions he wanted to turn over to his computer partner.

Licklider the psychologist drew a memorable comparison between a human and a computer. Humans are "noisy, narrow-band devices" but with many simultaneously active channels. Computer machines are fast and accurate, but they perform only a few elementary operations at a time. Humans are flexible, while computers are single-minded. Humans speak redundant languages while computers use a language with only two elementary symbols. To him, these contrasts suggest complementarity and symbiotic cooperation.

To fulfill his vision, Licklider calls for further development of time-sharing systems whereby a costly large-scale computer divides its time among many users. He also wants a "thinking center" with functions analogous to present-day libraries but with greatly enhanced information storage and retrieval. Then he imagines

> a network of such centers, connected to one another by side-band communications lines and to individual users by leased-wire services. . . . The cost of the gigantic memories and the sophisticated programs would then be divided by the number of users.[10]

Licklider thus anticipated two major research and development programs of ARPA that in fact did develop over the next decade: time sharing and computer information networks.

]]]]]] The Licklider Network

While Licklider is remembered as a highly original thinker, others in the Cambridge computing community shared his commitment to time sharing and interactive computing. After he moved to ARPA, he shaped a developing field not only because he generated intellectual excitement but also because he cultivated a network of a dozen of so pioneer computer scientists and engineers. Like him, most of them had gained experience with computers in the 1950s when they worked in institutions located in the Boston-Cambridge area. This environment provided a supportive context for early computer research and development in much the same way that the Los Angeles/Cal Tech environment had fostered early aerospace development.[11]

Many in Licklider's network had taken degrees in MIT's electrical engineering department, served as research assistants and associates in MIT's Digital Computer Laboratory, and/or held research posts in the computer division and groups at MIT's Lincoln Laboratory.[12] These facilities provided them access to several of the few large-scale research computers then available, including the early successors to the Whirlwind computer.

There were extensive shared institutional affiliations between the MIT/Lincoln Lab network and ARPA. Licklider, Ivan Sutherland, and Lawrence Roberts, three of the first four directors of the Information Processing Techniques Office of ARPA, had been MIT graduate students or faculty members and also researchers at Lincoln Laboratory. Leonard Kleinrock, whom we shall find playing a major role in ARPANET history, was an MIT classmate of Sutherland and Roberts.

At ARPA, Licklider moved into a strategic position that enabled him to champion his network of computer pioneers.[13] Under him and his immediate successor, IPTO became the country's leading source of research funds for computer development. He concentrated on selecting and funding research and development projects nominally categorized as military command and control–related but more precisely characterized as interactive computing, or man-machine symbiosis—his deepest commitment.

"In Cambridge everybody was excited about making it [interactive computing] exist," but so many developmental problems still remained that Licklider began searching for the best universities in which to fund interdisciplinary centers dedicated primarily to solving its problems. He had earlier selected graduate student assistants by canvassing outstanding universities and choosing the students who scored highest on the Miller analogy test. He believed that any who scored above 85 on the test should be hired, because a person with a gift for seeing analogies was sure to be "very good for something." Now he looked for those he considered to be the brightest computer scientists and engineers in these same universities. Asked how he identified the outstanding universities and people, he judged the question naïve. For him, finding outstanding people "is a kind of networking. You learn to trust certain people, and they expand your acquaintance. And the best people are in the best universities, which one knows by reputation."

Licklider obtained proposals from MIT, Carnegie-Mellon, the University of Utah, Stanford, UCLA, and the University of California–Berkeley, where he intended to fund centers of excellence. He identified principal investigators for projects by "going around and talking to people." He "got proposals out of" them rather than suggesting proposals to them. As a government funder, he perceived his role as responding to suggestions, not defining projects. A number of the suggestions came from alumni of MIT and Lincoln Laboratory who had fanned out to new computer professorships in research universities. He believed he could relate to these persons on a basis of trust rather than through a bureaucratic funding structure decorated with red tape. Even though he solicited ideas and proposals from computer scientists and engineers, Licklider's own overarching vision influenced the suggestions that they made. In his travels to centers of computer activity and to conferences, he deliberately talked about interactive computing and even about the possibility of a future network of computers. "We would get our gang together," Licklider recalls, "and there would be lots of discussion, and we would stay up late at night, and maybe drink a little alcohol and such." Conversation—and a little alcohol—often provoked responsive ideas, requests for funds, and gradually a loosely linked network of contracts—not unrelated to interactive computing. A system builder, Licklider conceived of the contracts as components from which coherent and systematic research and development programs that transcend individual universities could be fashioned.

His vision—his systems—not only responded to military needs, but also paved the high road to the future of nonmilitary computer engineering and science. He found that if he conceptualized his goals on a high level of abstraction—and if his principal investigators followed suit in similarly stating their objectives in their contract proposals—then "what the military needs is what the businessman needs is what the scientist needs." Their needs appear different only after the level of abstraction is reduced to specific tasks.

]]]]]] Time Sharing: MIT's Project MAC

The evolution in the computer field from a single computer serving a single user at a central location to time sharing, which provided access for a number of users with individual terminals to a single computer, is analogous to the evolution of electric light and power systems from isolated generating plants that supply electricity to a single household or commercial establishment to central generation stations that supply electricity by distribution lines to a number of consumers. The analogy is especially apt when applied to a time-sharing system with individual terminals connected by a telephone network to a central computer.[14]

Licklider considered time sharing a necessary preliminary to interactive computing. He argued that the introduction of time sharing in place of the current dependence on batch processing would greatly improve command and control systems. Needless to say, he saw time sharing as enhancing computing in the civil sector as well.

Licklider generously funded Project MAC (multiple-access computer or machine-aided cognition), a large, innovative time-sharing project at MIT. MAC not only provided increased and more convenient use of large mainframe computers but also became a major stepping-stone toward the networking of mainframe computers.

Licklider and Robert Fano, an MIT electrical engineering professor, conceived of Project MAC in 1962. Instead of seeking funding for a single MIT laboratory to develop time sharing, Fano submitted a multimillion-dollar proposal in January 1963 that would allow him to distribute funds among a number of MIT research centers that wanted to participate in the development of time sharing.

In his two-page request for funds, Fano argues:

> . . . the nation is facing many urgent information processing problems both military and civilian, such as those involved in command and control and information storage and retrieval. The number of people skilled in the techniques of information processing (not just programming) is insufficient to attack these problems both effectively and speedily.[15]

Project MAC, with an initial funding of $2.2 million, became operational in 1963. A year later the system served about two hundred users from a number of academic departments. At any one time, about thirty researchers could each use one of the hundred consoles, or teletype terminals, on campus or in faculty homes connected by telephone lines to an IBM mainframe computer. Despite a number of problems, such as the loss of centrally stored data, the installation and growing prominence of MAC encouraged a number of profit-motivated computer-utility ventures.[16] MIT professor John McCarthy, an early conceptualizer of time sharing, suggested by analogy the potential of commercialized time sharing. In a 1961 lecture, he predicted:

> If computers of the kind I have advocated [time sharing] become the computers of the future then computing may someday be organized as a public utility just as the telephone system is a public utility. . . . The computer utility could become the basis of a new and important industry.[17]

Commercial ventures began connecting paying customers to central mainframe computers possessing large stores of information and calculating power. In 1965, computer utilities and their stock issues became the "hottest new talk of the trade," but within several years serious difficulties in developing the needed software led to the bursting of the bubble and the folding of a number of companies that had sprung up to provide computer access for organizations and individuals. By 1970, computer utilities had become one of the "computer myths of the 1960s," thus providing one more example of a long history of false starts and dashed hopes in newly established, rapidly developing and changing fields of technology.[18] Several decades later, however, computer utilities flourished as hardware and software evolved.

On the other hand, Project MAC proved to be a fruitful learning experience for the research community using time sharing to access an interactive computer. Fano enthusiastically asserted that

> the availability of the MAC Computer System and of support for on-line interactive research resulted in a sudden explosion of computer research on the MIT campus. . . . On-line research with substantive external goals ranged from the development of problem-oriented languages in civil engineering to social systems analysis, from molecular model building to library information retrieval, from speech analysis to plasma physics, and from mathematical analysis to industrial dynamics.[19]

Time sharing suggested to a number of computer scientists the next critical problem to address on the advancing computer research and development front. Robert W. Taylor, successor to Licklider and Sutherland as head of IPTO, saw in a flash in 1966 the possibility of expanding the interactive community of computer users when he observed and speculated about the presence of three terminals in his Pentagon office. Each of these were connected by long-distance telephone lines to a time-sharing computer at an ARPA-funded research site. Because he often wished to communicate simultaneously with all three and because he shared Licklider's community-creating instincts, he decided that the three time-sharing computers should be interconnected to form a network.[20]

Like Licklider, Taylor possessed a vision of the future of computer communication, even though he, without hands-on computer experience, had to depend on others for the technical competence to fulfill the vision. Seeing that each of the various ARPA-funded sites "was digitally isolated from the other one," he decided "to build metacommunities by connecting them."

Precedent existed for such a network. Tom Marill, a psychologist who had studied under Licklider and who had established a small time-sharing utility, proposed to ARPA that he interconnect a Lincoln Laboratory computer with one in Santa Monica. ARPA counterproposed that Marill carry out the project under the aegis of Lincoln and that a young Lincoln researcher, Lawrence Roberts, later the ARPA administrator in charge of the ARPANET project, be in charge. Connected by Western Union lines, the computers exchanged messages, though the network was low in reliability and response time. Yet, Marill reported, he could "foresee no obstacle that a reasonable amount of diligence cannot be expected to overcome."[21]

]]]]]] Launching ARPANET

Licklider could take satisfaction in seeing Project MAC launched before he relinquished the reins of IPTO in 1964 to return to BBN. He had decided, however, that the time was not yet ripe for networking, time-sharing computers. He bequeathed this challenge to his successors. The successor IPTO heads carried on in his tradition of choosing and funding centers and projects. Sutherland, a Ph.D. in electrical engineering from MIT, followed Licklider (1964–66); Taylor, with an M.A. in psychology from the University of Texas, was next in line (1966–69)—we should recall that Licklider was also trained in psychology; Roberts, another MIT Ph.D. in electrical engineering followed (1969–73); then Licklider returned for another stint as IPTO head in 1974.[22]

Though Taylor came to ARPA from NASA, where he had been a program officer, the others came to ARPA for two- or three-year terms while they were on leave from a university or a research laboratory. Their circumscribed time at ARPA and their backgrounds contradict the oft-heard criticism that government funding of technology and science projects necessarily places control in the hands of time-serving, unimaginative bureaucrats. No bloated organization, IPTO remained lean, with a small staff of two or three assisting the head. IPTO's funding of a few large-figure projects rather than of many small ones reduced the administrative chores. The prospect of administering funding on a large scale seems to have been one attraction that brought capable administrators to IPTO.

Congress supported IPTO generously. Licklider and his successors learned from other ARPA administrators techniques best suited to approaching Congress. These included requesting funds for projects that had already demonstrated notable achievements; start-up funds for a project often came under the umbrella of a related project already under way.

IPTO heads found funding the development of a computer network congenial because it promised more than a 10 percent improvement in the state of the art; it was two or three years ahead of industry's achievements in the field; and it was a large-budget project likely to absorb a million dollars or more. During the first decade or so of the project, these funding criteria tended to be the norm at ARPA, especially at the IPTO.

IPTO initiated the ARPANET project in 1966 after IPTO head Taylor brought pressure on Lincoln Laboratory, an ARPA contractor, to release Roberts to preside as project manager over development of a computer network to interconnect time-sharing computers at the seventeen ARPA-funded academic, industrial, and government computer centers around the country. An analogy with the earlier history of electric power system development again becomes appropriate. Decades earlier, electric utility managers had begun to interconnect their central stations by transmission lines, thus forming regional power systems. The interconnections provided substantial economic advantages because a central station overloaded at any particular time could draw power from central stations in the regional system that were not overloaded. This raised the so-called load factor, or capacity utilization, of all the stations. A computer network could similarly allow sharing of resources to meet demand.

Roberts reluctantly left Lincoln, even though he had experienced difficulty in transferring his inventions and discoveries from the laboratory into the field. He recalls that their highly innovative work on networking computers remained largely unapplied. He believed that as an ARPA administrator he could act as a gatekeeper to spread ideas among university and industrial contractors. Military applications concerned him only remotely, even though he knew that ARPA projects had to interest the military.[23] ARPA also lured him because of his interest in interconnecting computers, an interest initiated by the same 1962 conference on the future of computing that had sparked Licklider's and Fano's interest in time sharing. Roberts remembers Licklider and "a bunch of people from MIT" talking late into the night about the future. A conversation about the need of researchers using computers in far-flung locations to be able to exchange data and software directly left a lasting impression on him.

Following Licklider's managerial style, Roberts and Taylor turned to their principal investigators, the computer scientists and engineers who presided over the ARPA-funded centers, for network algorithms, specifications, and means for performance evaluations. At an annual general meeting of principal investigators convened in Michigan in 1967, networking became the topic of discussion.[24] From that meeting, Roberts carried away several inventive concepts. Wesley Clark, an IPTO principal investigator at Washington University in St. Louis who had worked earlier on Project Whirlwind and SAGE at Lincoln Laboratories, suggested that host, or mainframe, computers should be interconnected not directly but through small interface computers that would provide a link between each host computer and the interconnecting network. Such an arrangement would permit host computers with different characteristics to connect through interface computers to a common communication network made up of telephone lines.[25] The role of the interface computers can be compared to that of the transformers, motor generator sets, and frequency changers that allow electric generating stations of different characteristics to connect to a transmission grid with its common voltage and frequency characteristics.[26] An alternative would have been to require all computers coming on the net to have standardized characteristics, but ARPA and its principal investigators preferred diversity.

While the principal investigators eagerly discussed the technical problems of creating a computer network, they proved far less enthusiastic once they had to face its funding requirements. ARPA saw networking as a means to reduce the principal investigators' need for funds to increase computer capacity and to support independent software development. Instead of purchasing increased capacity, ARPA wanted the seventeen computer centers to share capacity in a fashion similar to that employed by interconnected electric-generating stations. An overloaded station at any particular time, for example, would use the capacity of one underloaded at that period. At other times, the load circumstances of the stations might be reversed and the exchange would be in the opposite direction. This improved, as we have noted, each station's load factor.

The principal investigators initially preferred to have computer capacity completely under their local control and to develop their own software locally. After the net began operation, they began to see the advantages of resource-sharing. "It was only a couple years after they had gotten on it," Roberts remembers, "that they started raving about how they could now share research, and jointly publish papers, and do other things that they could never do before."

Within a few years, the concept of a computer network had become well enough established that in a 1970 article Roberts defined one as

> a set of autonomous, independent computer systems, intercon-
> nected so as to permit interactive resource sharing between any
> pair of systems. . . . The goal of the computer network is for each
> computer to make every local resource available to any computer
> in the net in such a way that any program available to local users
> can be used remotely. . . . The resources that can be shared in this
> way include software and data, as well as hardware.[27]

Independent computer "systems" referred to the time-sharing feature already in place at the various computer resource centers. Roberts foresaw that just as time-shared computer systems allowed hundreds of users to share hardware and software, networks inter-connecting dozens of time-sharing systems would permit resource-sharing by thousands. He did not predict that within a generation the number would be in the millions and that the network would become a commercial enterprise as well as a research facility.

]]]]]] Inventing ARPANET: Packet Switching

Besides the decision to develop small interface computers for the ARPANET, Taylor, Roberts, and their principal investigators decided to use packet switching. This technique had originally been proposed by Paul Baran, a researcher in the computer sciences department of RAND's mathematics division. Like other researchers who had earned a level of credibility from their RAND colleagues, Baran had considerable freedom in choosing his research problems. He decided that he could make the greatest contribution by tack-ling one aspect of a pressing Cold War problem upon which a num-ber of RAND researchers were working. They hoped to find ways of assuring the survival of a retaliatory strike force following a devastating attack. With a reasonable possibility of such a survival and retaliation, a first strike by either the Soviet Union or the United States became less likely. If there was no possibility of retaliation, either power might launch a peremptory strike, if the other gave evidence of preparing to launch an attack.

Baran chose to work on the problem of creating a communications network that could survive the strike and then command and control a retaliatory missile response. Not only did he believe that his experience with digital computers prepared him for such a task, but he also hoped that a secure communications network would alter the black-and-white choices of the Cold War adversaries to a gray—a light gray.[28]

After several false starts, he focused on the development of a "distributed adaptive message block network" system that he and several colleagues had generally defined by 1962 and then described in detail in a series of reports published in 1964.[29] "Distributed" refers to a noncentralized, nonhierarchical system in which the network transmission and reception nodes have a peer relationship. (By contrast, a hierarchical or centralized communication network has a single point of control from which connections extend to the various nodes. A single explosion could destroy the point of control.) The routing of messages is "adaptive" because they can be directed along different routes in the interconnected system to reach a particular node or destination. Each node is programmed to sense the availability of open connections or routing; if one route between nodes is loaded, then the node will automatically instruct message blocks to flow through alternative connections. "Message blocks" refers to the small packets into which messages are broken. The blocks are reassembled at their destination.

Baran later observed that it is relatively easy "to propose a global concept. It is far more difficult to provide enough details to overcome the hurdles raised by those that say 'It ain't gonna work.' " Stringent criticism came from the engineers in RAND's communications department who designed analog systems as well as from those engineers at AT&T who had long experience in designing and operating analog long-line telephone systems.

Baran found the AT&T headquarters people to whom he made his presentation gentlemanly—"Talk politely to them and they would invariably talk politely back to you"—but resolutely committed to analog systems and dismissive of digital ones. AT&T presided over a "monolithic" and "totally integrated" system requiring that any technological addition to the system had to fit in with existing equipment; it accepted evolutionary technological change, but rejected revolutionary change. Baran took as characteristic of the corporate culture a remark made to him by one AT&T engineer after an exasperating session: "First, it can't possibly work, and if it did, damned if we are going to allow creation of a competitor to ourselves." Later the creators of the ARPANET found AT&T unenthusiastic about providing the telephone connections they needed.

Even though RAND formally recommended to the Air Force that the distributed system be developed and despite the recommendation to proceed made by a 1966 Air Force evaluation review committee organized by MITRE, the system was not deployed until ARPA adopted and adapted it for the ARPANET several years later. Baran believes that the Defense Department did not have the in-house technical competence to develop the system.

Later reflecting about his contribution, Baran said:

The process of technological developments is like building a cathedral Over the course of several hundred years, new people come along and each lays down a block on top of the old foundations, each saying, "I built a cathedral." Next month another block is placed atop the previous one. Then comes along an historian who asks, "Well, who built the cathedral?" Peter added some stones here, and Paul added a few more. If you are not careful, you can con yourself into believing that you did *the* most important part. But the reality is that each contribution has to follow onto previous work. Everything is tied to everything else.

Baran's thoughts about the difficulty historians have in dealing with simultaneity of discovery and invention and about their cumulative nature are prescient. Historians may well decide that

Leonard Kleinrock—not Baran—through his research and publications introduced concepts later adapted for and embodied in the ARPANET as packet switching and distributed routing of data. When a Ph.D. candidate at MIT, Kleinrock did a dissertation, published in 1964, in which he analyzed the effectiveness of "time slicing," which anticipated packet switching, and of distributed control in data networks. Larry Roberts and others who designed the ARPANET were more familiar with Kleinrock's ideas than with Baran's.[30]

To add another stone to the "cathedral," at about the same time Donald Davies, at the National Physical Laboratory (NPL) in Teddington, England, also conceived of a version of message-block transmission that he called "packet switching." Without knowledge of Baran's earlier work, he sought to reduce the cost of using telephone lines as linkages for time-sharing computers. In the spring of 1966, Davies organized a seminar at the National Physical Laboratory where he presented his scheme for computer network communications. He referred then to message packets rather than message blocks. Financially constrained, Davies and associates at NPL then built a one-node network using packet switching, but on a much smaller scale than the ARPANET.

Baran, Kleinrock, and Davies constructed their systems on long-standing foundations built of the prior solid work of other pioneers. Telegraph stations had stored entire messages when lines were overloaded, forwarding them as the lines became free, much in the way that packet-switching nodes held and transmitted message blocks or packets. In the early 1960s, the military used a digital store-and-forward message-switching system called AUTODIN.[31]

]]]]]] Inventing ARPANET: IMP

The problem of interconnecting host computers of different manufacture and design was discussed at the annual principal investigators meeting in 1967. At that meeting, Wesley Clark suggested as a solution to the heterogeneity problem the placing of small interface, or gateway, minicomputers between host computers and the network. The host computers would see the interface computers as "black boxes" for providing an interface to the telephone line–linked network. The research centers would not need to be involved with, or even understand, the functioning of the "black-boxed" gateways, except insofar as it was necessary to design a hardware or software connection to them from the host computer. The separating and black-boxing of functional responsibilities became known as "layering." In this instance, the host computers constituted one layer and the gateway computers another. Layering reduced the complexity with which designers and operators had to contend.[32]

In 1967, Roberts found that his principal investigators judged the gateway scheme technically feasible, so he looked for an organization to design and build the projected gateway computers, which he called "interface message processors" (IMPs). Others referred to them as "packet switches" because they routed the message packets among alternative links interconnecting IMPs. In the summer of 1968, IPTO circulated to 140 potential bidders a "request for proposals" for the design and construction of a physical network including the IMPs and their software.[33] Drawn up by Roberts and principal investigators, the RFP provided a general conceptual design for the IMPs and also specified criteria that would be used to select the winning bidder. These included the understanding of the technical problems, the availability of experienced personnel, the performance characteristics of the small computer to be used as the IMP, and the general quality and commitment of the submitting firm to the project.[34] The RFP called initially for a four-node network to be deployed in nine months and a system capable in the long run of incorporating nineteen IMPs. Specifications called for IMPS to break messages into packets, to provide buffers to store packets, to route the message packets, and to monitor the network's flow of traffic.

Raytheon, Bunker-Ramo, Jacobi Systems, and Bolt Beranek and Newman, in addition to eight other firms, responded. In January 1969, IPTO awarded the contract to BBN, a small, entrepreneurial Cambridge, Massachusetts, research firm that Licklider served as vice president and whose early history we have summarized. BBN had previously received ARPA contracts and been designated one of the node sites on the projected computer network, but the IPTO contract amounted to an unusually large one for this organization of only several hundred people.[35]

BBN enjoyed a reputation as highly innovative, noted for its research and development. BBN proposed to depend on the nearby Honeywell company to provide a minicomputer that would be redesigned as an IMP and then to manufacture a number of IMPs. Because BBN researchers enjoyed considerable latitude in choosing their problems, the firm attracted outstanding young MIT and Harvard graduates who found front-edge research coupled with freedom from teaching a stimulating combination. In addition, generous contracts had provided the company excellent research and test facilities. In the eyes of IPTO, BBN's extensive experience in designing time-sharing networks also enhanced its credentials. The firm had established MEDINET, a medical system designed in collaboration with General Electric.

Robert Kahn, later an IPTO head but then an associate of the BBN team that developed the IMP, offered the following characterization:

> BBN was a kind of hybrid version of Harvard and MIT in the sense that most of the people there were either faculty or former faculty at either Harvard or MIT. If you've ever spent any time at either of those places, you would know what a unique kind of organization BBN was. A lot of the students at those places spent time at BBN. It was kind of like a super hyped-up version of the union of the two, except that you didn't have to worry about classes and teaching. You could just focus on research. It was sort of the cognac of the research business, very distilled. The culture at BBN at the time was to do interesting things and move on to the next interesting thing. There was more incentive to come up with interesting ideas and explore them than to try to capitalize on them once they had been developed.[36]

BBN organized a small team of five or six persons to design the interface message processor. The resulting set of design specifications embodied a number of inventive solutions to network problems. Future historians fully aware of the remarkable development of the worldwide Internet following hard upon the path-breaking ARPANET may some day compare the inventive success of the small BBN group to the achievements of Thomas Edison and his small band of associates who invented an electric lighting system.

Engineers and scientists with MIT and Lincoln Laboratory backgrounds dominated the BBN team. Frank Heart headed the team that drew up the proposal for ARPA and then developed the hardware and software. After obtaining a master's degree in electrical engineering from MIT, Heart had worked on the Whirlwind computer, then had been a researcher and group leader at Lincoln Laboratory for fifteen years before moving to BBN in 1966. William Crowther, who had the responsibility for developing the IMP software, had a master's degree in physics from MIT; he had been a researcher at Lincoln Laboratory for ten years before he went to BBN. Severo Ornstein, designer of the IMP hardware, had a Harvard undergraduate degree in geology and had learned about computers in the field, having served as a researcher at Lincoln Laboratory for seven years. David Walden, who assisted Crowther with the software, after taking an undergraduate degree in math from San Francisco State College, then did research at Lincoln Laboratory for three years. Kahn contributed to the team as an authority on communications theory and as a designer of systems architecture. Earlier, he had earned a Ph.D. in electrical engineering from Princeton University before becoming an MIT faculty member in electrical engineering and a researcher on MIT's Project MAC.

Heart characterizes his team's endeavor as "a labor of love"; members of the team refer to their work as "fun." He encouraged team members to think holistically about the project and to reach decisions by consensus. Only as a last resort did he employ his authority as team leader.[37] Walden remembers that "mostly what happens is you sit in a room and argue about it until you all agree about what the right answer is."[38] Heart's style, like that at the IPTO office at the time, involved finding "bright" people interested in the problem at hand and giving them free rein.[39] The BBN culture called for macro-, not micro-, management.

Many of the engineers and scientists engaged in the ARPANET project began their professional careers during the turbulent counterculture 1960s. They became enthusiastic advocates of consensus over hierarchical management. They stressed the meritocratic nature of their problem-solving communities, which were also populated mainly by keen and dedicated scientists and engineers. The work characteristics of several of the BBN team members reinforce the supposition that the counterculture values of the 1960s influenced their behavior. Heart on one occasion expressed concern that Crowther would wear his sneakers to a high-level meeting at ARPA headquarters in Washington, D.C., which in fact he did—but without disturbing the tenor of the proceedings. Ornstein, who took part in antiwar demonstrations, only half-jokingly threatened to pin a resistance button on the colonel with whom the BBN team was negotiating.[40] In their recollections, members of the team generally describe their relations with the military as remote, despite their being a military agency and the ARPANET's having originated with military needs. Ornstein recalls that the team members felt "insulated" from the military.

Heart's team had offices side by side so that they could engage in countless informal meetings to discuss design problems. They kept informal working memoranda called "the IMP Guys' notes." (In time, the BBN team began to refer to itself as the IMP Guys.) When a breakthrough occurred, Walden remembers, "We'd run in and say, 'Look, I got this running. Somebody come and type on the teletype. . . . This is exciting. Something is cycling.' "[41]

The designers of software and hardware closely interacted. Ornstein, who had principal responsibility for developing the hardware, remembers solving problems by spending countless hours late into the night at his home with Kahn, the expert on systems architecture. Kahn had previously had little experience with designing hardware, but his omnivorous interest in all aspects of the IMP project led him to question Ornstein incessantly, thereby learning in depth from

THE ORIGINAL IMP GUYS: TRUETT THATCH, BILL BARTELL (HONEYWELL), DAVE WALDEN, JIM GEISMAN, BOB KAHN, FRANK HEART, BEN BARKER (BEHIND HEART), MARTY THORPE, WILL CROWTHER, SEVERO ORNSTEIN, BERNIE COSELL (NOT PICTURED, AND HAWLEY RISING (NOT PICTURED). *(Courtesy of Frank Heart)*

him. Ornstein himself worked closely with Crowther, the software expert, whom he found to be "a brilliant programmer . . . [who] thoroughly understood machine language code, the kind of code that you have to whittle down."[42]

Walden recalls that designing the IMP was essentially a problem in engineering design rather than the application of theory:

> It becomes an engineering problem as opposed to a theory problem. . . . We had to send these bits down the wire: how do you put a header on the front; how do you put a trailer on the back. There was a theory of how you put error correcting codes on it. Bob Kahn knew that theory and told us what it was. There were some constraints: this is the way that the 303 (or the 301 or

whatever the Bell modem is) has to be interfaced to, but after that it was all pretty pragmatic. Not lots of theory coming from someplace else.[43]

The group designing the IMP did not rely heavily on information, or communications, theory. The BBN team, like the engineers and scientists who designed the heat shields for the Atlas and Titan missiles, could find little theory relevant for guiding their empirical thrusts, explaining their empirical successes, or rationalizing their empirical designs. Walden recalls that what was later "taught in courses in communications about networks and protocols and all of that, I would say we were mainly . . . inventing it—the academic analysis tended to come later."

Kahn, who had studied information theory and worked at Bell Laboratories, where Claude Shannon, the leading theoretician, had developed his concepts, often questioned the engineering empirical, or go-ahead, approach of other members of the team, some of whom always had their heads "right down in the bits." He applied theoretical analysis insofar as he could in designing simulations of message traffic flows predicted for the network. His associates found that he moved along the mountaintops of theory, but often did not have the patience to explain the details of an idea that he was championing. Expecting them to grasp his ideas, he might say, "Don't you see it? It is all there."[44] Because his approach differed sharply from theirs, some of the other members of the team despaired of his ever becoming a "computer person" despite his learning from Ornstein. They felt that he would "never come to understand the problems looking at them his way."[45] In time, the team found that had they heeded some of the criticisms from Kahn and others of a like-minded theoretical bent, it would have saved them some missteps. Furthermore, the team could have drawn more heavily on prior network theory published by Kleinrock and others. Kleinrock believes that the BBN team concentrated too much on obtaining an IMP design that would work and not sufficiently on developing a network that would perform well under various constraints.[46]

During the nine months that elapsed between the awarding of the contract and the delivering of the first IMP, Ornstein and his assistants concentrated on adapting a Honeywell 516 computer for use as an IMP. Crowther and his associates wrote the software for it. Ornstein found working with the Honeywell to adapt its 516 computer for use as an IMP trying. He had assumed that the 516 was a mature machine, but he discovered "bugs" in it and had to "diddle" with it extensively. He characterized the Honeywell engineers with whom he dealt as being "industrial strength," not "research strength" people. Most of the time, Honeywell sent "cabbages instead of computers"; the company not only delivered machines behind schedule but ones that did not work. Ornstein had to become "quite nasty at times and beat on the table." Yet Honeywell shaped up, producing special hardware under great pressure.[47]

Crowther strove for software that would enable the IMP to route the packets among the alternative routes to their destinations in such a way as to minimize cost and time of transmission and to optimally utilize the capacity of the distributed network. To do this, the software provided each IMP information continuously about the state of traffic, or information, flows throughout the network. Furthermore, when a route was not instantly available, the software placed message packets in waiting queues, or buffers. Crowther, who considers designing a routing algorithm a "fun" thing to do, tells us about his approach:

> If given a complex system and an algorithm, like a routing algorithm, I tend to be pretty good at visualizing the thing and seeing what will happen and what some of the bad cases are. So there were a lot of mental things like that. When you came up with one that looked pretty good, then you'd try it and see whether or not it worked.[48]

Shortly, we shall consider the testing done to "see whether or not it worked."

LEONARD KLEINROCK, AN ARPANET PIONEER. *(Courtesy of Leonard Kleinrock)*

ⅠⅠⅠⅠⅠⅠ The Network Working Group

The BBN team delivered the first IMP in September 1969 to Leonard Kleinrock's computer research center at the University of California–Los Angeles. The ARPA office had chosen this site as the initial node on the four-node experimental ARPANET. In short order, Walden, the junior software designer, delivered three more to the University of California–Santa Barbara, the Stanford Research Institute (SRI), and the University of Utah.

Kleinrock, one of ARPA's well-funded principal investigators, had a group of about forty working in his center: secretaries, programmers, managers, faculty, and graduate students. He had studied in MIT's electrical engineering department, then served on the Lincoln Laboratory research staff at the same time as Roberts. They, with Sutherland, who preceded Roberts as head of IPTO, stood their final MIT Ph.D. theses defenses together in 1959. Though they worked on independent projects, all three of them used Lincoln's TX-2 computer. One of the most complex computers available at the time, a successor to the Whirlwind computer, TX-2 was a device on which many pioneers in the computer field learned computing, hands-on. Kleinrock used the TX-2 to simulate computer networks, an experience that prepared him for his contribution to the development of the ARPANET. In 1963, he left the Lincoln Laboratory to take a professorship at UCLA where he discovered that "I love teaching, I love research, I love the environment."[49]

When Kleinrock watched the first IMP being wheeled into his computer research center, he and his team faced the daunting tasks of installing both hardware and software that would allow his center's host computer to "talk to," or "interface with," its IMP. The team could expect some help from Kahn, of the BBN team, who had spent weeks drawing general specifications for the connection between a host and its IMP.[50] While BBN took full responsibility for connections and communications among IMPs, Kleinrock's graduate students had the responsibility for developing the host-to-host protocol that would enable the UCLA host computer to communicate with host computers at other sites, or nodes, on the ARPANET.[51]

Writing the host-to-host "protocol" program proved to be one of the most difficult software problems encountered in deploying ARPANET. Traditionally, protocol refers to an agreement among diplomats concerning the etiquette and precedence that will facilitate communication, deliberation, and negotiation. Analogously, in the world of computer networks, protocol, besides designating agreed-upon format, syntax, and semantics of messages, decides upon the signaling information. This information is appended to a message that directs the movement of the message from sender to recipient. In similar fashion, the address on a conventional letter provides the signal generating the activity that results in the delivery of the message by the post office.[52]

In 1969, Kleinrock and other principal investigators set up the Network Working Group (NWG), a committee that included graduate students, to decide upon protocols for the ARPANET, especially those to interconnect host computers. Roberts assigned

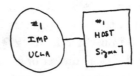

THE ARPA NETWORK

SEPT 1969

I NODE

THE ARPA NETWORK, SEPTEMBER 1969. *(Courtesy Computer Museum History Center)*

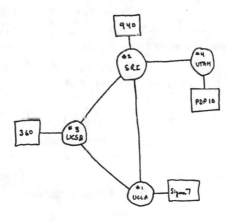

THE ARPA NETWORK

DEC 1969

4 NODES

THE ARPA NETWORK, DECEMBER 1969. *(Courtesy Computer Museum History Center)*

the protocol problem to a committee because he believed that ingenious ideas were dispersed among the members of the various ARPA-funded computer research centers and because he wanted all the potential sites on the ARPANET to have a stake in the project. A disproportionate number of graduate students from the computer research centers served on the NWG. In retrospect, this is not surprising because they were solving many of the computer problems that their professors had previously identified and defined.

A UCLA graduate student at the time, Vinton Cerf, who later became a major figure in the computing field, recalls:

> We were just rank amateurs, and we were expecting that some authority would finally come along and say, "Here's how we are going to do it." And nobody ever came along, so we were sort of tentatively feeling our way into how we could go about getting the software up and running. In the longer term, Larry Roberts was very insistent that this intrepid bunch of graduate students—not just at UCLA but at other sites like MIT, and Utah, and SRI, and UC Santa Barbara—get their rear ends in gear and actually make decisions about the protocols and get them instantiated and get them running in all the operating systems.[53]

Another UCLA graduate student and member of the UCLA Computer Club, which was made up of young computer enthusiasts, Steve Crocker became de facto head of the Network Working Group and its spokesperson at meetings held mostly at ARPANET sites on the East and West coasts. Crocker also initiated the taking of minutes of the meetings. These he circulated with requests for comments (RFC). Through the minutes, other notes, and the responses to RFCs, the Network Working Group community accumulated a large store of experience-based information about the processes of designing protocols. Roberts did not insist that the committee solve problems in a particular way, only that the members avoid the academic tendency to haggle over fine points and, instead, to reach by consensus decisions that could be tested empir-

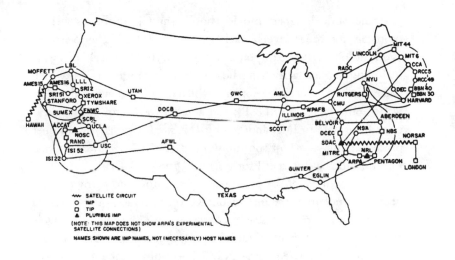

THE ARPANET, JUNE 1977. *(Courtesy Computer Museum History Center)*

ically. Alexander McKenzie, who took part in NWG deliberations as the BBN representative, suggests the management style of Roberts and ARPA:

> The program managers at ARPA didn't really have the time to tell people how to do it. They really much more aimed at funding people who were smart enough and self-motivated enough to recognize what the problems were and go solve them. There was, in my experience, zero or close to zero micro-management of any ARPA program—either at BBN or anywhere else that I ever heard of. ARPA's objectives were to find people that they thought were sufficiently smart and sufficiently motivated and give them a ball and let them run with it.[54]

The committee decided to define protocols in layers. Layers, as we have indicated, are analogous to the levels of abstraction used in language studies. An example is the move up the ladder of abstraction from bricks, to walls, to rooms, and finally to houses. Architects provide a general plan for the house as well as detailed plans for walls and rooms. The house is one layer, the walls another. Similarly, computer network designers provide host-to-host protocols as well as protocols for other layers, such as host-to-IMP.[55] The Network Working Group settled upon basic ARPANET protocols in 1972, including one for facilitating host-to-host communications, which had taken two years to develop and which became known as the Network Control Program (NCP); and another to control communications from host to IMP and among IMPs.[56] In the 1970s, another protocol for host-to-host communication, the TCP/IP (Transmission Control Protocol/Internet Protocol), replaced the NCP.

Because most of the principal investigators and many of their graduate students at the intended ARPANET nodes—to be located at academic computer research centers—initially took a dim view of Roberts's plan to bring their host computers onto the ARPANET, haggling characterized some NWG meetings. Not only did the footdraggers prefer having additional resources under their unilateral control to sharing them across the network, but they also realized that adapting their host computers to the protocol software meant that "they would have to do major surgery on their operating system to get this network to talk to them."[57] Responding to this resistance, the designers of the protocols kept them simple so as to minimize the difficulty of writing the program for adapting the host computers to the requirements of the common protocols. Roberts persisted in building the network by brandishing carrots and sticks until, by August 1972, the fifteen host-computer sites connected to the network were using the NCP protocol.

Heart, as IMP team manager at BBN, found that

> it was a surprise how tough it was to agree on the host protocols. . . . The network's utility was delayed for at least two years because of a misapprehension of how hard it was going to be to get the host protocol suites in place. . . . It's like picking up the phone and calling France—if you don't speak French you've got a little problem. So even if you get the connection to the two telephones, if you don't speak French you don't communicate very well . . . I think it was misperceived how long and hard that would be to do.

And he adds:

> While they might have some differences, the overriding single
> bit was they were all having a great time. And they all thought it
> was very exciting.

]]]]]] The Network Measurement Center: Testing

We have noted the critical feedback links between design and testing during the development of the Atlas missiles. The conceptual
design of a technological system and its components consists often
of a set of hypotheses about what might work, what might fill the
system's requirements, or goals. Testing is a means of validating or
invalidating the hypotheses. If tests fail, the designers modify the
hypothesis. Alternatively, the testing may show that the failure
resulted from prototypes used in testing not having been constructed in accord with the design specifications. The initial deployment of the four-node ARPANET amounted to a research prototype
needful of testing.

The Network Measurement Center funded by ARPA and
headed by Kleinrock did experimental design and stress testing.
His earlier research on queuing theory and its application to evaluating computer network performance prepared him well for the
task. Having written his dissertation on store-and-forward networking, he had a theoretical understanding of the functioning of
the IMPs and of the communications network of which they were a
part. In drawing up the preliminary specifications for the network
and the IMPs in 1968, Roberts had turned to Kleinrock for advice.

Kleinrock and the graduate students created theoretical models of projected network performance (including failures such as
transmission deadlock). They decided how to test the actual performance of the ARPANET and to compare this performance with an
idealized network. Cerf, as a UCLA graduate student, assumed a
leading role in the testing, working closely with Kahn of BBN,
who shared Kleinrock's interest in theory and testing. Earlier, Kahn
had tried to persuade his colleagues at BBN, including project

leader Heart, that the BBN team had done insufficient theoretical analysis and testing before shipping the initial IMPs.[58] Roberts, often in contact with Kahn, tended to agree. He not only assigned the Network Measurement Center the testing project, but also gave a contract for network topology studies to the Network Analysis Corporation, a small for-profit firm headed by Howard Frank that specialized in network design analysis.[59] Frank, an electrical engineer from the University of California–Berkeley, and a friend of Kleinrock's, had made a name for himself several years earlier when he and Kleinrock analyzed and reorganized the layout of a major pipeline system, thereby saving the operators sizable operating expenses.

Encouraged by Kahn, a Kleinrock graduate student wrote software that loaded, or drove, heavy message traffic into the network in order to discover how many packets would be lost and the length of transmission delays that would result from such overloading. Kleinrock and Kahn observed the behavior of the congestion-control mechanism on the IMPs that decided when to switch packet routes. For example, packets could travel directly from UCLA up to the Stanford Research Institute, or alternatively they could move to SRI by way of the IMP at Santa Barbara. On occasions an IMP switched back and forth rapidly between two routes looking for the less congested path. If the paths were almost equally loaded and the IMP added traffic to one, then this route became the congested one, so the IMP switched to the other, and so on, back and forth.

Kahn had worried about the reliability of Crowther's design for the software that controlled the IMPs dynamic routing of messages through the IMP subnet. Kahn predicted that congestion on a heavily loaded subnet would occasionally cause a system "lockup," or bottleneck, because the storage capacity of the IMPs would be filled to capacity with message packets, thus additional incoming packets would be rejected. Katie Hafner and Matthew Lyon in their history of the ARPANET, *Where Wizards Stay Up Late* (1996),[60] describe "grand little fights" among the BBN team members about the design of the congestion-control mechanisms. Ornstein said that "some of the things . . . [Kahn suggested] were off the wall, just wrong." Gradually, the Heart team paid less attention to Kahn's strictures: "Most of the group were trying to get Kahn out of our hair," Ornstein recalls.

In January 1970, with the four-node network in place, Kahn decided to visit Kleinrock's UCLA center in order to test his theory that the network could have a congestive failure. He took Walden with him to manipulate the IMP code so as to vary the size and frequency of the packets passing over the network from IMP to IMP. "By besieging the IMPs with packets, within a few minutes he and Walden were able to force the network into catatonia."[61] After returning to BBN, Kahn showed results from this and other tests to Heart and Crowther. Finally persuaded that the network could lock up, Crowther worked with Kahn to rectify the problem. Kahn could argue, in retrospect, that earlier theory should have guided practice; the IMP Guys could argue that theory was playing its proper role of following upon and rationalizing practice.

]]]]]] The 1972 Demonstration

With ARPA regularly adding computer research centers to the network, Roberts wanted to increase net traffic. In the fall of 1971, the network ran at only about 2 percent capacity and only the inner sancta of the computer research community knew of the network's existence.[62] The slow development of network protocols partially explains the low utilization of the network. Believing that more information about the potential of the network needed to be disseminated, Roberts encouraged publication of articles about the ARPANET in professional journals. He also decided to sponsor a public demonstration of the new technology, a practice often resorted to by inventor-entrepreneurs. At the turn of the century, wireless and airplane inventors had repeatedly used public demonstrations to raise capital and to sell their patented devices.

Roberts asked Kahn in 1971 to organize a demonstration for the first International Conference on Computer Communication to be held in Washington, D.C., at the Hilton Hotel in October 1972.

By then, the Network Working Group had defined additional protocols and BBN had adapted the IMPs so that a number of remote telephone-line-connected dial-in user terminals could be linked to a host computer in a manner reminiscent of time sharing. In addition, a Network Control Center, directed at that time by Alex McKenzie, a young Stanford-trained computer engineer at BBN, and staffed by several persons at BBN, monitored the performance of ARPANET and did remote troubleshooting.

Kahn recalls that the demonstration made the ARPANET "real to others . . . people could now see that packet switching would really work. It was almost like the train industry disbelieving that airplanes could really fly until they actually saw one in flight." Over a thousand persons interested in networking watched as computers of different manufacture, operating in the display room in a Washington hotel, communicated with other computers located at various sites on the ARPANET. Kahn and his associates urged visitors to use terminals themselves to log in to various host computers, exchange data and files. Most of the present and future leaders in the networking field were on hand. "It was a major event. It was a happening," Kahn concludes.[63]

The 1972 demonstration in conjunction with the availability of protocols and the increased reliability of the ARPANET changed the image of the ARPANET. Computer engineers and scientists no longer considered it a research site for testing computer communications but saw it as a communications utility comparable to that of the telephone system. "It was remarkable how quickly all of the sites really began to want to view the network as a utility rather than as a research project," McKenzie confirms. He and the Network Control Center wanted the ARPANET to perform as reliably as an electric power utility, but he acknowledges that in the early years the IMPs were up only 98 or 99 percent of the time, which "would be an abysmal record for a power utility."[64]

The unanticipatedly heavy use of electronic mail (e-mail), especially for personal messages as contrasted with professional ones, also moved ARPANET down the utility path. As early as 1973, e-mail constituted three-quarters of ARPANET traffic. Not intended by its developers to be a message system, the ARPANET by the end of the 1970s nonetheless derived its greatest stimulus for growth from the e-mail traffic.[65] The history of e-mail provides substantial support for those who stress the unintended consequences of invention and development.

]]]]]] The Internet

Our emphasis has been upon the creation of the ARPANET, not upon its post-1972 deployment and transformation as an operating utility. We can, however, summarize some of the milestones in its postinnovational history. In 1975, ARPA, in accord with its policy of turning over research and development projects to the military once the projects had become operational, transferred the management of the network to the Defense Communications Agency, which manages communications for the military. As a result, military needs increasingly shaped the further course of ARPANET.[66]

ARPA, however, continued to fund computer network research. In 1973, Kahn, who had become an administrator at IPTO and its head from 1979 to 1985, and Cerf, who began teaching at Stanford University in 1972 and in 1976 became a program manager at IPTO, together conceived of the basic architecture of an "internet" that would interconnect ARPANET with several packet-switching networks that ARPA had established after 1970. One network used radio and another used satellites, instead of telephone lines, to provide the communication subnet.

Kahn and Cerf faced the problem of interconnecting networks with differing characteristics, a problem similar to that of interconnecting different kinds of host computers on the ARPANET. Once again focusing upon the problem of defining protocols, Cerf and Kahn published a paper in 1974 in which they defined the TCP/IP protocol for use in sending messages across network interfaces. Gateway computers, placed at network interconnection points, function not unlike the motor-generator units that made possible interconnecting alternating-current and direct-current electric power networks early in the twentieth century.

Competition soon developed among those organizations favoring TCP/IP and those choosing an OSI (open systems interconnection) protocol defined by the International Organization for Standardization. In 1983, the Department of Defense helped resolve the conflict by requiring all host computers on the ARPANET to use TCP/IP. When other networks, including those located abroad, made the transition, an international internet, now known as the "Internet," came into being. In 1983, the Defense Department divided the ARPANET, which by then interconnected hundreds of host computer sites, into a smaller ARPANET and a MILNET, the former dedicated to the computer research community and the latter to military users.[67]

The National Science Foundation funded five supercomputer centers in 1986. These became the backbone of the NSFNET (National Science Foundation Net), which was used extensively by computer researchers in universities. With former ARPANET functions largely taken over by public, commercial, and private networks, including the NSFNET, the Defense Department ended ARPANET's existence as a distinct system. The Internet continued to flourish along with the establishment of an increasing number of networks that interconnected through gateway computers and the TCP/IP protocol. During the 1990s, the creation of the World Wide Web and the introduction of browsers, or search engines, have stimulated the extensive use of the Internet by individuals and profit-seeking organizations.

]]]]]] Management of ARPANET

In our histories of SAGE and Atlas, we stressed management of the research and development process as we focused on the upper layers of a large managerial structure that presided over thousands of contractors. Only occasionally did we catch a glimpse of the way in which a small team of ten to twenty scientists and engineers organized the management of a small-scale problem-solving project. By

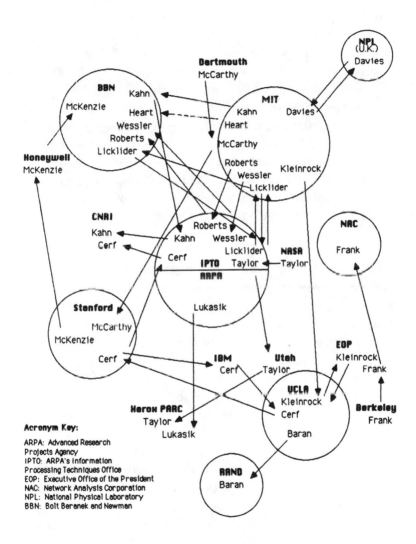

DYNAMIC NETWORK OF THE ARPANET PIONEERS. THE ARROWS INDICATE MOVEMENT FROM ONE ORGANIZATION TO ANOTHER. *(Courtesy of Jane Abbate)*

contrast, the history of ARPANET has directed our attention to the managerial creativity of small teams of computer scientists and engineers, such as the one at BBN, the Network Measurement Center, and the small ARPA/IPTO management team.

The language and concepts used by ARPANET engineers, scientists, and managers to describe the processes in which they were involved recall earlier episodes in the history of technology such as that of Thomas Edison and his team's invention and development of an electric lighting system, Elmer Sperry and his small team's invention and development of the airplane stabilizer, and Henry Ford and his team's invention and development of the moving assembly line.[68] The style of the ARPANET engineers, scientists, and managers also brings to mind projects of other computer research and development teams as described in Tracy Kidder's *The Soul of a New Machine* (1981) and G. Pascal Zachary's *Showstopper!* (1994).

Heart of BBN, for example, stresses that persons such as Roberts, who performed the management role for IPTO, possessed technical competence—they were "not just managers." Because of their technical backgrounds, the "not just managers" preferred, Heart believes, to discuss and arbitrate rather than to dictate when differences of opinion obtruded. They assumed that scientists and engineers, unlike persons more politically motivated, would ultimately recognize "right" reasoning instead of simply deferring to those with hierarchical authority. He concedes, however, that Roberts often participated in consensus forming with a "louder voice" than the others, in part because he controlled the funds.[69]

While Heart and others speak approvingly of consensus, they did not take an egalitarian approach to problem solving. They highly valued the ideas of people whom they considered bright and who usually came from elite university backgrounds, especially MIT. McKenzie of BBN describes ARPA's style as giving "bright people" the authority to choose other "smart" people and give them freedom to do research, emphasizing results rather than costs.[70] We should also note that ARPA/IPTO referred to computer research centers as centers of excellence, placing them in major research universities. Much as ARPA/IPTO encouraged its self-motivated and informed researchers to reach decisions by consensus, it similarly viewed the research centers as horizontally related, with authority and responsibility distributed among them. ARPA/IPTO considered hierarchical structure taboo. Not even the ARPANET was centrally controlled. We should recall the importance that Roberts, in deciding on protocols, attached to the regular deliberative meetings of the principal investigators from the computer research centers as well as his resort to the Network Working Group.

The ARPA management sustained a "collegial," even an academic, environment at the research centers. It brought scientists and engineers, on leave from research universities, to Washington for several years to manage IPTO. These university scientists and engineers had either had experience in managing research and development or had observed the techniques firsthand in working with entrepreneurial professors and researchers at institutions such as MIT. Maintaining close ties to universities, IPTO directors Licklider, Sutherland, Taylor, and Roberts kept abreast of front-edge academic research. They imbued IPTO with traditional values of academic research, such as freedom of inquiry and dissemination of information.

In the case of large projects such as SAGE and Atlas, managers motivated the engineers and scientists by stressing the national defense imperatives driving the mission-oriented project. While Roberts had to persuade several of the principal investigators that it was to their advantage to connect to the ARPANET, most of the computer scientists and engineers designing and developing the network seem to have been self-motivated by the "fun" of problem solving and by the satisfaction of working on the frontier and advancing the exciting new field of computing and communications. IPTO managers rarely spoke to the scientists and engineers of a military mission and never of possible commercial objectives. Heart sums up a widespread spirit of self-motivation:

You know the people who were involved in this all were having a very good time. The ARPANET was a big thing in most of their lives. So most of the groups of the host sites, or the Network Analysis Corporation, or at Kleinrock's Network Measurement Center, or here, or others, were all having a very good time. They were all really having the time of their lives. While they might have some differences, the overriding single bit was they were all having a great time.

]]]]]] Large Contracts, Bright People

ARPA attracted technically competent and highly motivated managers by assuring them that not only could they award large contracts but that they would have a reasonably free hand in doing so. So encouraged, IPTO managers translated Licklider's emphasis on quality into a policy of nurturing a dozen or so "centers of excellence" by funding large projects, or sets of projects, at a few locations so as to concentrate expertise. Besides focusing on centers of excellence, the IPTO managers placed contracts in the hands of those whom they labeled "bright people." As we have observed, Taylor, following Licklider's lead, found those who fulfilled his criteria by questioning the large network of computer scientists and engineers with whom he regularly talked.[71] He and the other IPTO heads depended on the principal investigators at research centers and others in the computer community to feed in project proposals. But at the same time, the IPTO managers stimulated the flow of ideas by raising questions and presenting problems at the annual meetings of the principal investigators and through conversations with the computer community throughout the country.

Taylor awarded sustaining multiyear contracts to research centers. MIT, for instance, received about $3 million a year, Carnegie-Mellon several million annually, and Stanford and UCLA hundreds of thousands. He tried to award contracts to projects that fell somewhere between efforts to define "the nature of God and to count the grains of sand on the beach." Projects that had a high chance of success and that would make an order of magnitude of difference caught his attention.

Like other project managers, Roberts had to coordinate and schedule, but he managed only a handful of contractors while Atlas's Schriever and his staff dealt with hundreds. He did not have either a large control room to display contract progress reports or weekly critical-problems meetings as did Schriever. Roberts was able to monitor, schedule, and coordinate by making his own mental syntheses and analyses. Besides coordinating the ARPANET activities of BBN, the Network Measurement Center at UCLA, and the Network Analysis Corporation, Roberts also had responsibility for contracts awarded the Stanford Research Institute. This institute maintained a Network Information Center, gathering and providing information to the network community about the characteristics and performance of the ARPANET. Roberts coordinated the activities of the contractors by informal site visits during which he learned about progress being made, obstacles encountered, and additional resources needed.

]]]]]] Past and Present

Like so many of the managers, engineers, and scientists who played major roles in the development of government-funded projects between 1950 and 1970, ARPANET principals look back years later to a heroic period in their professional lives when they participated in the early projects. McKenzie of BBN is representative in observing:

> And I know that there has been fraud and abuse in government and in contracting and so forth, but it seems to me that the kind of rules and regulations that there are now, that are attempting to prevent that, really make it very difficult for the government to get the same kind of power out of its research dollars these days as it was able to then. I know it's hard to find a balance between accountability and free rein, and these days the government

approach seems to be more on the side of accountability and less on the side of free rein. But I think that a lot is being lost. . . . I think that ARPA in the 1970s did a really good job for the country in that way. It was a joy to be associated with the ARPANET project. It was fun. It was challenging. And I think it was good for the country. It's not so easy to find that mix now, and I think regulation is a big part of it.

VII Epilogue:

Presiding over Change

▼ During the half-century following World War II, America continued to produce a cornucopia of material goods through modern management and engineering. Alongside this capitalistic, free-enterprise achievement, the country's capacity to create the large-scale technological systems that structure our living spaces has grown as well. Post–World War II government-funded projects have introduced a creative management and engineering style substantially different from one called modern that flourished during the period between the two world wars.

The modern, or pre–World War II, managerial and engineering approach associates management with large manufacturing firms rather than with joint ventures and projects such as SAGE, Atlas, Central Artery/Tunnel, and ARPANET. Unlike post–World War II managers and engineers involved with projects that introduced new technological systems, such as computer networks and urban highways, the prewar modern managers and engineers became employees of well-established firms whose products changed only incrementally.

The maintenance of a system for mass-producing standardized products engaged prewar modern engineers and managers. On the other hand, the postwar engineers and managers whom we encountered are not rigidly committed to standardization. They tolerate, even embrace, heterogeneity. Managers and engineers in the modern period expected that the problem-solving techniques they had mastered as young professionals would change only marginally. By contrast, post–World War II project engineers and managers need frequent refresher courses to keep abreast of the rapidly changing state of the art. Linear growth became the hallmark of modern management and engineering; discontinuous change is the expectation of project professionals. The modern professionals assumed research and development to be a prelude to an enduring innovation. Project engineers and managers now realize that managing innovation is the norm and anticipate a sequence of research and development projects.

A typical modern pre–World War II manufacturing firm was a multiunit enterprise, such as the General Motors Corporation with its various divisions, including Chevrolet and Buick, managed by a hierarchy of mid- and top-level executives. Even though they were divided into units, General Motors and similar firms were tightly coupled by an overarching, hierarchical management layer. The numerous contractors participating in projects like SAGE and Atlas are loosely coupled by information networks and by a coordinating and scheduling systems engineering organization such as Ramo-Wooldridge.

Yesterday's engineers and managers endeavored to tightly integrate an organization or a production system. The scientific management philosophy of Frederick W. Taylor and his followers and the continuous-production flow doctrine of Henry Ford encouraged this approach. SAGE and Atlas demonstrate the capability of systems engineers to masterfully coordinate heterogeneous technologies and autonomous contractors into a loosely coupled system in order to obtain a clearly defined goal.

The advantages of the earlier form of business enterprise could not be achieved "until a management hierarchy had been created. . . . the existence of a managerial hierarchy is a defining characteristic of the modern business enterprise." This hierarchy became a source of power, continued growth, and "permanence."[1] A modern firm, thus organized, depended upon programmed, or directive, control while project management depends upon feedback from its loosely coupled contractors.

Because recent projects often concentrate on research and development, hierarchy has to be modified to allow for local initiative. The resulting compromise, called "black-boxing," allows local research and development teams to choose the technology that will fulfill system specifications. When, for example, systems engineers specify the thrust required from a missile propulsion engine, the propulsion design team has leeway in choosing the mechanical, electrical, and chemical means for achieving these specifications. The systems engineers need not look into the "black box" within which the team is working, and micromanage the means used to achieve the specified ends.[2]

Pre–World War II manufacturing firms valued the highly trained engineering specialists who concentrated on mechanical, electrical, or chemical problems. But postwar projects foster an interdisciplinary approach that cultivates the generalist. Modern experts long familiar with a firm's product line expected deference from young engineers and managers. Today, experience counts for less on projects with a large research and development component because the problems encountered are often unprecedented. Projects sustain a meritocracy.

]]]]]] Postmodern Projects

While the SAGE, Atlas, CA/T, and ARPANET projects share a number of characteristics unlike those of earlier management and engineering, SAGE and Atlas differ from the other two in several ways. They are transitional projects demonstrating both modern (prewar) and postmodern (postwar) characteristics. The management and engineering that created the Central Artery/Tunnel and the ARPANET, however, differ so much from the modern that they can be characterized as essentially postmodern.[3]

The postmodernity of CA/T and ARPANET can be attributed in no small part to the counterculture values that spread in the 1960s. Many Americans blamed the tragedy of the Vietnam War on their government's amoral use of large-scale military technology. This distrust of government and technology focused not only on military technology but on large-scale technology in general. Hierarchy, bureaucracy, and governance by experts fell into disfavor.

The systems approach that created a regional air defense system and an intercontinental ballistic missile also lost credibility. It proved unable to cope with the messy problems that arose in the housing, transportation, health, and poverty programs of President Johnson's Great Society. Atlas and SAGE focused primarily on technical and economic factors; such matters as the environment, political-interest-group commitments, and public participatory-design concerns lay beyond the horizons of 1950s systems engineering.

Unlike SAGE and Atlas managers and engineers, those involved in the Central Artery/Tunnel Project take into account the concerns of environmental and interest groups. Through public hearings, the project fosters participatory design. CA/T is not an elegantly reductionist endeavor; it is a messily complex embracing of contradictions.

The CA/T style of management and engineering results in large part from the vociferous and effective demonstrations in the 1960s and 1970s against urban highway construction. Only after the organizers of CA/T made clear that they would take into account public concerns about neighborhood integrity and the environment was the project funded. CA/T has been socially constructed, not technologically and economically determined.

The ARPANET Project also displays characteristics associated with the counterculture. Creators of ARPANET preferred a flat management structure as contrasted with the vertical one exemplified by General Motors. Interactions among ARPANET engineers and scientists were collegial and meritocratic. ARPA/IPTO project managers encouraged self-motivated and informed researchers to reach decisions by consensus. They viewed the research centers they funded as horizontally related, with authority and responsibility distributed among them. ARPA/IPTO considered hierarchical structures taboo.

Characteristics of project and postmodern technology and management can be contrasted with characteristics of modern technology and management.[4] The polarities that follow are too sharply drawn to accord with reality, but the contrasts suggest the differing management and engineering styles.

MODERN	POSTMODERN
production system	project
hierarchical/vertical	flat/layered/horizontal
specialization	interdisciplinarity
integration	coordination
rational order	messy complexity
standardization/homogeneity	heterogeneity
centralized control	distributed control
manufacturing firm	joint venture
experts	meritocracy
tightly coupled system	networked system
unchanging	continuous change
micromanagement	black-boxing
hierarchical decision making	consensus-reaching
seamless web	network with nodes
tightly coupled	loosely coupled
programmed control	feedback control
bureaucratic structure	collegial community
Taylorism	systems engineering
mass production	batch production
maintenance	construction
incremental	discontinuous
closed	open

Acknowledgments

During a concentrated research and writing project extending over six years and culminating in the publication of this book, my indebtedness to friends, colleagues, and organizations has grown enormously. The full commitment to the project came when Harriet Zuckerman of the Andrew W. Mellon Foundation offered to generously fund a project of my choice provided the research, writing, and publications would contribute to the understanding of the history of technology broadly conceived and to the training of graduate students. Her unflagging support and encouragement has continued throughout the project.

Dan Frank, editorial director of Pantheon Books, has also nurtured my endeavor. Having been deeply impressed by his impeccable editorial taste and his editing of my most recent book, *American Genesis,* I wanted him to care for this one as well. His patient and insightful support throughout the years of book gestation has deeply impressed me again. His editorial assistant, Claudine O'Hearn, has been faithfully supportive as well.

Because of the Mellon support, I have worked with a spirited and talented group of young scholars who acted as research associates at various times during the project. As I name them, good memories of a cooperative endeavor flood in: Janet Abbate, Atsushi Akera. Nathan Collins, Karin Ellison, Elliot Fishman, Glenn Bugos, David Foster, Fred Quivik, Erik Rau, and Amy Slayton. I am especially grateful to Julie Johnson of Harvard University, who presided over the acquisition of the illustrations. A friend of my wife's, Julie volunteered to do this time-consuming work in her memory.

Besides the scholarly support of young professionals, I had the logistical assistance of dedicated administrators at both the University of Pennsylvania and the Massachusetts Institute of Technology. I wish to acknowledge especially the support at Penn of Neal Hébert and Hannah Poole, who administered the Mellon funds, and of Debbie Meinbresse, who cheerfully lent a helping hand at MIT.

From research seminars also come thematic concepts and information useful in making books. I was especially stimulated by the discourse in extended seminars on large technological systems at the University of Pennsylvania, MIT, the Royal Institute of Technology in Stockholm, and Rathenau Summer Academies.

Agatha Hughes and I have worked side by side throughout this project. Longtime friends who are also professional colleagues supported us in manifold ways over the years. I speak for her and for me as I fondly recall professional and deeply personal relationships with Mary Anderson, Nancy Bauer, Stephen Bauer, Denise Scott Brown, Bernard Carlson, Paul Edwards, Yehuda Elkana, Gabrielle Hecht, Sheldon Isakoff, Julie Johnson, Arne Kaijser, Dan Kevles, Tim Lenoir, Svante Lindqvist, Donald MacKenzie, Everett Mendelsohn, Joachim Nettelbeck, Johannes Ottow, Glenn Porter, Cheri Ross, M. Roe Smith, Anne Spirn, John Staudenmaier, Jane Summerton, Georg Thurn, Robert Venturi, and Rosalind Williams.

Professional colleagues have also contributed substantially to the project. Among them are Jed Buchwald, Carla Chrisfield, Lorraine Daston, Davis Dyer, Loren Graham, David Hounshell, Bernward Joerges, Philip Khoury, Todd M. LaPorte, Joel Moses, David Noble, Arthur Norberg, Ted Postel, Skuli Sigurdsson, Gene Rochlin, Dan Roos, Robert H. Roy, Evelyn Simha, M. Roe Smith, Nils-Eric Svensson, David Warsh, Sarah Wermiel, and G. Pascal Zachary.

Several persons reviewed and commented on chapters in rough draft. For this time-consuming contribution, I thank Janet Abbate, Jay Forrester, Loren Graham, Katie Hafner, Frank Heart, Lucian Hughes, Richard John, Leonard Kleinrock, Peter Markle, Arthur Norberg, Simon Ramo, Fred Salvucci, Bernard Schriever, Thomas Smith, and a group of engineers and managers at the Central Artery/Tunnel Project.

In the Notes, I have acknowledged numerous interviews

which have been enormously helpful in writing about recent history. Other persons gave interviews that provided a general overview. Not able to make reference to them at specific locations in the text, I acknowledge their helpful interviews here. They are Alex Almeda, Gwen Bell, Robert Bressler, Harvey Brooks, Matt Coogan, Bernard Frieden, Jr., Nikolas Hoff, Michael Jellinik, Richard John, Carl Kaysen, Peter Markel, Matthew Meselson, Robert Mitchell, Walter Rosenblith, Jack Ruina, Harvey Sapolsky, Tom Smith, Gene Solnikov, Ivan Sutherland, Edward Teller, Stan Weiss, Jerry Welch, and Robert Wood.

While interviews are essential for the writing of recent history, archives remain bedrock support for historians. Among those I used and the archivists who assisted me are Raymond Puffer of the Air Force Ballistic Missile Organization, Bruce Bruemmer of the Charles Babbage Institute, Judith Goodstein of the California Institute of Technology, Patty Flannagan, David Kruh, and Holly Sutherland of the Central Artery/Tunnel Project, David Baldwin of MITRE Corporation, and Helen W. Samuels and Elizabeth Andrews of the MIT Archives. Zoe Allison provided ARPANET illustrations, and Daniel J. McNichol gave invaluable aid in acquiring CA/T illustrations.

I could not have sustained the effort and invested the thought and energy in an endeavor for which the light at the end of the tunnel was often not visible without a loving family, which includes my daughter, Agatha Heritage Hughes, and my son, Lucian Parke Hughes, and his family.

My beloved wife Agatha suddenly and unexpectedly died on 14 July 1997, one week after we completed the manuscript. Throughout almost fifty years she and I loved deeply and did history together.

Notes

I. INTRODUCTION: TECHNOLOGY'S NATION

1. In *American Genesis* (New York: Penguin, 1990), an account of earlier American history, I followed the development of large-scale technology from 1880 to about 1940, an era dominated by independent inventors and industrial-research-laboratory scientists.
2. This book discusses large technological, or engineering and applied science, projects. Since World War II, there have been many large science projects, especially ones creating large scientific instruments. See Peter Galison and Bruce Hevly, eds., *Big Science: The Growth of Large-Scale Research* (Stanford, Calif.: Stanford University Press, 1992). In this volume, see, for example, Allan Needel, "From Military Research to Big Science: Lloyd Berkner and Science—Statesmanship in the Postwar Era," pp. 290–311.
3. For a discussion of the design process, see Louis Bucciarelli, *Designing Engineers* (Cambridge, Mass.: MIT Press, 1994).
4. The "universities" in the complex often manage large research laboratories that depend on military funding. Michael A. Dennis, "A Change of State: The Political Cultures of Technical Practice at the MIT Instrumentation Laboratory and the Johns Hopkins University Applied Physics Laboratory, 1930–1945" (Ph.D. diss., Johns Hopkins University, 1991).
5. I refer to the "Atlas Project," but Simon Ramo, the organizer of systems engineering for the project, reminds me that Atlas was only the first of three coupled intercontinental ballistic missile projects of the 1950s and 1960s. He prefers to refer to the "ICBM Project."
6. General Bernard A. Schriever, "The USAF Ballistic Missile Program," in *The United States Air Force Report on the Ballistic Missile*, ed. Kenneth Gantz (New York: Doubleday, 1958), p. 30.
7. For an analysis of the visual thinking and creativity of inventors and engineers, see W. Bernard Carlson and Michael E. Gorman, "A Cognitive Framework to Understand Technological Creativity:

Bell, Edison, and the Telephone," in *Inventive Minds,* ed. R. Weber and D. Perkins (New York: Oxford University Press, 1992), pp. 48–79. See also Michel Callon and John Law, "Engineering and Sociology in a Military Aircraft Project: A Network Analysis of Technological Change," *Social Problems* 35 (January 1988): 284–97.

8. David McCullough, *The Path Between the Seas: The Creation of the Panama Canal, 1870–1914* (New York: Simon & Schuster, 1977); Joseph E. Stevens, *Hoover Dam: An American Adventure* (Norman: University of Oklahoma Press, 1988).

9. Richard Rhodes, *The Making of the Atom Bomb* (New York: Simon & Schuster, 1986); Richard G. Hewlett and Oscar E. Anderson, Jr., *A History of the United States Atomic Energy Commission: The New World, 1939–1946* (University Park: Pennsylvania State University Press, 1962). On the Radiation Laboratory, see Robert Buderi, *The Invention That Changed the World: How a Small Group of Radar Pioneers Won the Second World War and Launched a Technological Revolution* (New York: Simon & Schuster, 1996).

10. See especially Bruno Latour, *Aramis; or, The Love of Technology* (Cambridge, Mass.: Harvard University Press, 1996); Tracy Kidder, *The Soul of a New Machine* (New York: Avon, 1990); G. Pascal Zachary, *Showstopper!* (New York: Free Press, 1994); Glenn Bugos, *Engineering the F-4 Phantom II: Parts into Systems* (Annapolis, Md.: Naval Institute Press, 1996); Craig Canine, *Dream Reaper: The Story of an Old-Fashioned Inventor in the High-Stakes World of Modern Agriculture* (New York: Knopf, 1995); Katie Hafner and Matthew Lyon, *Where Wizards Stay Up Late: The Origins of the Internet* (New York: Simon & Schuster, 1996); Douglas K. Smith and Robert C. Alexander, *Fumbling the Future: How Xerox Invented, Then Ignored, the First Personal Computer* (New York: Morrow, 1988); and Fred Moody, *I Sing the Body Electronic* (New York: Viking, 1995).

11. Timothy Lenoir, *Instituting Science: The Cultural Production of Scientific Disciplines* (Stanford, Calif.: Stanford University Press, 1997), p. 240. Besides the numerous and informative books in the NASA-sponsored history series, see Walter A. McDougall, *The Heavens and the Earth: A Political History of the Space Age* (New York: Basic Books, 1985).

12. Robert Kanigel, *The One Best Way: Frederick Winslow Taylor and the Enigma of Efficiency* (New York: Viking, 1997); Stephen P. Waring, *Taylorism Transformed: Scientific Management Theory Since 1945* (Chapel Hill: University of North Carolina Press, 1991).

13. Robert M. White, "The Migration of Know-how," *The Bridge* 24 (Winter 1994): 11.

14. John N. Warfield and J. Douglas Hill, *A Unified Systems Engineering Concept*, Battelle Monographs, ed. Benjamin Gordon (Columbus, Ohio: Battelle, 1972), p. 1. The complex nature of the systems approach to management differs starkly from the concept of management presented in many "how-to" books now circulating among the public and aspiring managers. The system builders and managers encountered here did not, however, discount the psychological and sociological traits commonly associated with leadership. They had to combine leadership and expertise in order to cope with complexity.

15. Merritt Roe Smith, ed., *Military Enterprise and Technological Change: Perspectives on the American Experience* (Cambridge, Mass.: MIT Press, 1985); David A. Hounshell, *From the American System to Mass Production, 1800–1932: The Development of Manufacturing Technology in the United States* (Baltimore: Johns Hopkins University Press, 1984).

16. Thomas Parke Hughes, *Elmer Sperry: Inventor and Engineer* (Baltimore: Johns Hopkins University Press, 1971).

17. For a penetrating critique of the military-industrial complex, see Seymour Melman, *Pentagon Capitalism: The Political Economy of War* (New York: McGraw-Hill, 1970); and Carroll W. Pursell, Jr., ed., *The Military-Industrial Complex* (New York: Harper & Row, 1972).

18. Melman, *Pentagon Capitalism;* Mary Kaldor, *Baroque Technology* (New York: Hill & Wang, 1981); Pursell, *The Military-Industrial Complex.*

19. For a poignant account of the demise of the defense engineers' blue-sky dream, see David Beers, *Blue Sky Dream: A Memoir of America's Fall from Grace* (New York: Doubleday, 1996).

20. Alain Enthoven letter to the editor, [Stanford] *Reporter,* 2 June 1995.

21. *New York Times,* 23 August 1953. The summary of its editorial position is taken from a survey of editorials from 1949 to 1953 done by Max Collins, a graduate student research assistant on this book project.

22. For other interpretations of the relationship between the military and academia, see Stuart Leslie's thoughtful study *The Cold War and American Science* (New York: Columbia University Press, 1992); and Paul Forman and José M. Sanchez-Ron, eds., *National Military*

Establishments and the Advancement of Science and Technology (Norwell, Mass.: Kluwer, 1996).

23. Beers, *Blue Sky Dream.*

24. Professor Ted Postel of MIT called this to my attention.

25. Simon Ramo, *The Business of Science: Winning and Losing in the High-Tech Age* (New York: Hill & Wang, 1988).

26. MIT political scientist Harvey Sapolsky observes that in the 1990s "the rationales that garner the most support for science are no longer national security rationales, but rather health and economic growth," rationales "much less protective of its [science's] independence." Harvey M. Sapolsky, *Science and the Navy: The History of the Office of Naval Research* (Princeton, N.J.: Princeton University Press, 1990), p. 127.

27. On the other hand, the role of labor in the technological process does not emerge from our history of CA/T, as it does not in the accounts of the other projects, because our focus is on design and development, not construction.

28. For more on public participation, see Brian Balogh, *Chain Reaction: Expert Debate and Public Participation in American Commercial Nuclear Power, 1945–1975* (New York: Cambridge University Press, 1991).

29. Bruce E. Seely, *Building the American Highway System: Engineers as Policy Makers* (Philadelphia: Temple University Press, 1987).

II. MIT AS SYSTEM BUILDER: SAGE

1. Karl T. Compton to N. M. Sage, 8 September 1948, copy in author's possession.

2. Report L-3 from J. Forrester and staff to N. Sage with cover letter from N. M. Sage to K. T. Compton, 17 September 1948, copy in author's possession.

3. Robert Buderi, *The Invention That Changed the World: How a Small Group of Radar Pioneers Won the Second World War and Launched a Technological Revolution* (New York: Simon & Schuster, 1996), p. 358.

4. Outstanding among the members were Charles S. Draper, an SAB member who headed the MIT Instrumentation Laboratory; H. Guyford Stever, also a SAB member from MIT and an aeronautical engineer who later became a presidential science adviser; and John Marchetti, civilian director of the Air Force Cambridge Research Laboratories (AFCRL). Dr. W. Hawthorne of MIT was listed as assisting the committee on aircraft propulsion problems.

5. Kent C. Redmond and Thomas M. Smith, *Project Whirlwind: The History of a Pioneer Computer* (Bedford, Mass.: Digital Equipment Corp., 1980), p. 172.

6. Charles J. Smith, *History of the Electronic Systems Division, SAGE: Background and Origins,* vol. 1, AFSC Historical Publication Series, 65-30-1 (Bedford, Mass.: Historical Division, Office of Information, Electronic Systems Division, Laurence G. Hanscom Field, Air Force Systems Command, 1964), pp. 68–69.

7. Buderi, *The Invention That Changed the World,* p. 363.

8. Smith, *History of the Electronic Systems Division,* pp. 69, 71.

9. James R. Bright, "The Development of Automation," in *Technology in Western Civilization,* ed. Melvin Kranzberg and Carroll W. Pursell, Jr. (New York: Oxford University Press, 1967), pp. 2, 635–37.

10. Air Defense Systems Engineering Committee, "Air Defense System: ADSEC Final Report," 24 October 1950, MITRE Corporation Archives, Bedford, Mass., pp. 2–3.

11. Ibid., p. 10.

12. George E. Valley, "How the SAGE Development Began," *Annals of the History of Computing* 7 (July 1985). 211.

13. Jack S. Goldstein, *A Different Sort of Time: The Life of Jerrold R. Zacharias: Scientist, Engineer, Educator* (Cambridge, Mass.: MIT Press, 1992), p. 115.

14. *SAGE: Semi-Automatic Ground Environment,* a color videotape of a conversation involving several persons, including Jerrold Zacharias, Jay Forrester, and James Killian, who played leading roles in the project. Prepared under the auspices of the Alfred P. Sloan Foundation. Arthur Singer, Sloan vice president, called this historic document to my attention.

15. J. R. Marvin and F. J. Weyl, "The Summer Study," *Naval Research Reviews,* 20 August 1966, pp. 1–7; James R. Killian, Jr., *The Education of a College President: A Memoir* (Cambridge, Mass.: MIT Press, 1985), pp. 71–72.

16. Office of the Chairman of the Corporation, "Government Supported Research at M.I.T.: An Historical Survey Beginning with World War II: The Origins of the Instrumentation and Lincoln Laboratories," MIT, 1969, MIT Institute Archives, p. 27.

17. Goldstein, *A Different Sort of Time,* pp. 93, 106.

18. Buderi, *The Invention That Changed the World,* p. 375.

19. Goldstein, *A Different Sort of Time,* p. 99.

20. There have been, however, many other claims about the origins of systems engineering in the Navy and in the military in general.

21. Miss R. Joyce Harman was the only woman listed on the committee. She probably served as a writer and editor.

22. Project Charles, *Problems of Air Defense: Final Report of Project Charles,* Massachusetts Institute of Technology internal document, 1 August 1951, MIT Institute Archives, pp. 3,5.

23. Paul N. Edwards, "The World in a Machine: Origins and Impacts of Early Computerized Global Systems Models" (paper presented at the "Spread of the Systems Approach" conference held at the Dibner Institute, MIT, Cambridge, Mass., May 1996), p. 16.

24. Killian, *The Education of a College President,* p. 72.

25. Goldstein, *A Different Sort of Time,* p. 111.

26. John F. Jacobs, *The Sage Air Defense System* (Bedford, Mass.: MITRE Corp., 1986), p. 17. We should also recall that other members of the Research Laboratory of Electronics had criticized the recommendations of the Valley committee. For an informed survey account of SAGE in the context of Cold War military strategy, see Paul Edwards, *The Closed World: Computers and the Politics of Discourse in Cold War America* (Cambridge, Mass.: MIT Press, 1996), pp. 75–111.

27. Letter of transmittal for *Final Report of Project Charles,* quoted in Richard F. McMullen, *The Birth of SAGE, 1951–1958,* Air Defense Research Command Historical Study no. 33, n.d. For an interesting discussion of recent thinking about command and control and military electronics, see Manuel de Landa, *War in the Age of Intelligent Machines* (New York: Swerve Editions, 1991).

28. Jay W. Forrester, "From the Ranch to System Dynamics," in *Management Laureates: A Collection of Autobiographical Essays,* ed. Arthur G. Bedeian (Greenwich, Conn.: JAI Press, 1992), p. 340. This is volume 1 in a series.

29. The scale and scope of the Whirlwind project grew not only because of Forrester's expanding vision, but also because other persons and organizations found it in their interest to commit resources to the project. Bruno Latour argues that "every time a new group becomes interested in the project it transforms the project." He calls this kind of development the "whirlwind" or "translation" model of innovation. *Aramis; or, The Love of Technology* (Cambridge, Mass.: Harvard University Press, 1996), pp. 118–19.

30. Redmond and Smith, *Project Whirlwind,* pp. 127–28.

31. Ibid., p. 101.

32. Valley, "How the SAGE Development Began," pp. 209–10.

33. Buderi, *The Invention That Changed the World,* p. 38.
34. Ibid., p. 369.
35. Valley, "How the SAGE Development Began," p. 210.
36. Testimony of Jay W. Forrester, "Jay W. Forrester vs. Jan A. Rajchman: Interference No. 88,269," United States Patent Office, Examiner of Interferences, 1960, MIT Institute Archives.
37. Testimony of Jay W. Forrester, pp. 23–25; quotation, p. 27.
38. Valley, "How the SAGE Development Began," p. 213.
39. For more on Sperry, see Thomas Parke Hughes, *Elmer Sperry: Inventor and Engineer* (Baltimore: Johns Hopkins University Press, 1971; paperback ed., 1994).
40. Forrester, "From the Ranch to System Dynamics," p. 337.
41. Redmond and Smith, *Project Whirlwind,* p. 183.
42. Ibid.
43. Ibid., p. 38.
44. Ibid., p. 181.
45. Valley, "How the SAGE Development Began," p. 217.
46. I am indebted to Elliot Fishman, former graduate student at the University of Pennsylvania, for several informative research reports on the magnetic core memory. He drew on the essays and books of Emerson Pugh, including *Memories That Shaped an Industry: Decisions Leading to IBM System/360* (Cambridge, Mass.: MIT Press, 1984); and Charles J. Bashe et al., *IBM's Early Computers* (Cambridge, Mass.: MIT Press, 1986).
47. Forrester, "From the Ranch to System Dynamics," p. 342.
48. For the subsequent history of the development of the computer as a control device in industry, see David W. Noble, *Forces of Production: A Social History of Industrial Automation* (New York: Knopf, 1984); and J. Francis Reintjes, *Numerical Control: Making a New Technology* (New York: Oxford University Press, 1991).
49. John Jacobs, "SAGE Overview," *Annals of the History of Computing* 5 (October 1983): 325.
50. Wheeler Loomis to James R. Killian, Jr., 21 December 1951; James R. Killian to Thomas K. Finletter, 21 December 1951; and Finletter to Killian, 5 February 1952, MIT Institute Archives.
51. James R. Killian, Jr., to Thomas K. Finletter, 9 January 1953, MIT Institute Archives.
52. Jacobs, *The Sage Air Defense System,* p. 26.
53. Thomas K. Finletter to James R. Killian, Jr., 15 January 1953, MIT Institute Archives.

54. E. E. Partridge to James R. Killian, 28 January 1953, MIT Institute Archives. Partridge dropped the "Jr." from Killian's name, but bestowed upon him the title of "Dr.," which he was not.

55. Smith, *History of the Electronic Systems Division,* p. 91.

56. E. E. Partridge to James R. Killian, 28 January 1953, MIT Institute Archives.

57. Valley, "How the SAGE Development Began," p. 221.

58. Redmond and Smith, *Project Whirlwind,* p. 203.

59. E. E. Partridge to James R. Killian, 6 May 1953, MIT Institute Archives. See also McMullen, *Birth of SAGE,* p. 19.

60. Redmond and Smith, *Project Whirlwind,* p. 203.

61. Valley, "How the SAGE Development Began," p. 222.

62. Ibid., p. 220.

63. Ibid.

64. Author's interview with Jay Forrester at MIT, 20 May 1994.

65. Redmond and Smith, *Project Whirlwind,* p. 203.

66. Jacobs, *The Sage Air Defense System,* p. 22.

67. Valley, "How the SAGE Development Began," pp. 196–226.

68. Ibid., pp. 224. Perry Goode headed the Michigan project, but Valley does not identify him as the Michigan representative.

69. Jay W. Forrester to A. G. Hill, 12 May 1953, MIT Institute Archives.

70. Jay W. Forrester to A. G. Hill, 12 May 1953; A. G. Hill to Raymond C. Maude, 13 May 1953, MIT Institute Archives.

71. On IBM's contribution to SAGE, see Emerson Pugh, *Rebuilding IBM: Shaping an Industry and Its Technology* (Cambridge, Mass.: MIT Press, 1995), pp. 207–19.

72. D. L. Putt to James R. Killian, 17 November, 1954, MIT Institute Archives.

73. Pugh, *Memories That Shaped an Industry,* p. 96.

74. Author's interview with Jay Forrester at MIT, February 1992.

75. Pugh, *Memories That Shaped an Industry,* p. 97.

76. Ibid., p. 126.

77. Ibid.

78. Technical Memorandum no. 20, "A Proposal for Air Defense System Evolution: The Transition Phase," first draft, 1 December 1952; second draft, 2 December 1953, MITRE Archives.

79. Headquarters, Air Defense Command, Ent Air Force Base, Colorado Springs, Colorado, *Operational Plan: Semiautomatic Ground Environment System for Air Defense,* 1955, MITRE Archives.

80. Before the circulation of the "Red Book," the Lincoln Transition System had become known as SAGE.

81. *Operational Plan,* p. 79.

82. Thomas Petzinger Jr., "History of Software Begins with the Work of Some Brainy Women," *Wall Street Journal,* November 15, 1996. The article is based on research done by Kathryn Kleiman for a video documentary.

83. For the history of System Development Corporation, see Claude Baum, *The System Builders: The Story of SDC* (Santa Monica, Calif.: System Development Corp., 1981).

84. Herbert D. Benington, "Production of Large Computer Programs," *Annals of the History of Computing* 5 (October 1983): 350.

85. Ibid., p. 351; Donald MacKenzie, "A Worm in the Bud? Computers, Systems, and the Safety-Case Problem" (paper presented at the "Spread of the Systems Approach" conference held at the Dibner Institute, MIT, Cambridge, Mass., May 1996), p. 6.

86. Glenn Bugos, "Programming the American Aerospace Industry, 1954–1964: The Business Structures and the Technical Transactions," *Journal of the Business Studies Ryukoku University* 35 (June 1995): 65–76.

87. R. R. Everett to C. F. J. Overhage and W. H. Radford on "Air Defense Systems Engineering," 3 June 1957, MITRE Archives.

88. Jay W. Forrester to M. G. Holloway, 7 March 1956, MIT Institute Archives.

89. Author's interview with Robert Everett at MITRE headquarters, 11 October 1994.

90. Buderi, *The Invention That Changed the World,* p. 376.

91. Author's interview with Robert Everett.

92. Kent C. Redmond and Thomas M. Smith, "The R&D Story of the SAGE Computer," unpublished manuscript, 1994, MITRE Archives, pp. 23–27.

93. R. R. Everett to C. F. J. Overhage and W. H. Radford on "Air Defense Systems Engineering," 3 June 1957, MITRE Archives.

94. Ibid.

95. Jacobs, *The Sage Air Defense System,* p. 138.

96. Jacobs, "SAGE Overview," pp. 327–28.

97. Jacobs, *The Sage Air Defense System,* p. 139.

98. Ibid., p. 141.

99. McCormack, who named MITRE, declared this was no acronym, but only a word connoting the new organization's charac-

ter. On the history of MITRE, see Robert Meisel and John F. Jacobs, *MITRE: The First Twenty Years: A History of the MITRE Corporation (1958–1978)* (Bedford, Mass.: MITRE Corp., 1979).

100. David Alan Rosenberg, "The Origins of Overkill: Nuclear Weapons and American Strategy, 1945–1960," *International Security* 7 (Spring 1983): 32, 38, 46–47. See also Marc Trachtenberg, *History and Strategy* (Princeton, N.J.: Princeton University Press, 1991).

101. Kenneth Schaffel, *The Emerging Shield: The Air Force and the Evolution of Continental Air Defense, 1945–1960* (Washington, D.C.: Office of Air Force History, United States Air Force, 1991), p. 197.

102. Author's interview with Robert Everett at MITRE headquarters, 11 October 1994.

103. Author's interview with Charles Zraket at MITRE headquarters, 15 January 1992.

104. Edwards, "The World in a Machine," p. 18. On the management of weapons and warfare, see Manuel de Landa, *War in the Age of Intelligent Machines.*

105. John F. Jacobs, *Practical Evaluation of Existing Highly Automated Command and Control Systems* (Bedford, Mass.: MITRE Corp., 1965).

106. Arthur Norberg, interview with Charles Zraket, Charles Babbage Institute, Minneapolis, 1990, transcript OH 198.

107. Ibid.

108. Ken Olsen, the founder of the Digital Equipment Corporation, was a Whirlwind alumnus.

109. "Return on Investment in Basic Research—Exploring a Methodology: Report to the Office of Naval Research, Department of the Navy," Bruce S. Old Associates, 1991.

110. Robert R. Everett, "Whirlwind Genesis and Descendants" (a talk given at the Computer Museum, Boston, Massachusetts, 18 October 1987), copy in author's possession.

111. Statement of M. O. Kappler, president, Systems Development Corporation, in U.S. Congress, House of Representatives, Committee on Government Operations, Military Operations Subcommittee, *Systems Development and Management,* pt. 3, pp. 989ff. The hearings were in Washington, D.C., during the period from 21 June to 15 August 1962.

III. MANAGING A MILITARY-INDUSTRIAL COMPLEX: ATLAS

1. "Armed Forces: The Bird and the Watcher," *Time*, 1 April 1957, p. 17. Simon Ramo recalls twenty principal contractors.

2. On the German missiles, see Michael Neufeld, "The Guided Missile and the Third Reich: Peenemünde and the Forging of a Technological Revolution," in *Science, Technology, and National Socialism*, ed. Monika Renneberg and Mark Walker (Cambridge: Cambridge University Press, 1990).

3. Ann Markusen et al., *The Rise of the Gunbelt: The Military Remapping of Industrial America* (New York: Oxford University Press, 1991), pp. 91–92.

4. Ibid., p. 91.

5. Gene Marine, "Think Factory De Luxe," *The Nation*, 14 February 1959), pp. 131–35.

6. Fred M. Kaplan, "Scientists at War: The Birth of the Rand Corporation," *American Heritage* 34 (June–July 1983): 52.

7. Theodore von Kármán, *The Wind and Beyond* (Boston: Little, Brown, 1967), p. 271.

8. Edmund Beard, *Developing the ICBM: A Study in Bureaucratic Politics* (New York: Columbia University Press, 1976), p. 24.

9. Author's interview with Francis Clauser at California Institute of Technology, 14 February 1992; author's interview with Homer Joseph Steward in Altadena, California, 13 February 1992.

10. Jacob Neufeld, *The Development of Ballistic Missiles in the United States Air Force 1945–1960* (Washington, D.C.: Office of Air Force History, United States Air Force, 1990), pp. 30–36.

11. Ibid., pp. 44–45.

12. John L. Chapman, *Atlas: The Story of a Missile* (New York: Harper & Brothers, 1960), p. 28.

13. Ibid., p. 32.

14. Richard E. Martin, "The Atlas and Centaur 'Steel Balloon' Tanks: A Legacy of Karel Bossart" in *Proceedings of the 40th International Astronautical Congress* (Málaga, Spain: American Institute of Aeronautics and Astronautics, 1989), p. 14.

15. Chapman, *Atlas*, pp. 32, 33.

16. For more on the guidance and control of missiles, see Donald MacKenzie, *Inventing Accuracy: A Historical Sociology of Nuclear Missile Guidance* (Cambridge, Mass.: MIT Press, 1991).

17. Chapman, *Atlas*, p. 59.

18. Neufeld, *The Development of Ballistic Missiles*, p. 48.

19. Chapman, *Atlas*, pp. 36–53; Neufeld, *The Development of Ballistic Missiles*, pp. 44–50.
20. Beard, *Developing the ICBM*, p. 62.
21. Ernest G. Schwiebert, "USAF's Ballistic Missiles—1954–1964," *Air Force/Space Digest*, May 1964, p. 67.
22. Beard, *Developing the ICBM*, pp. 58–61.
23. Ibid., p. 82.
24. Robert L. Perry, *The Ballistic Missile Decision* (Santa Monica, Calif.: RAND Corp., 1967), pp. 26–27; Perry, "Commentary on I. B. Holley's 'The Evolution of Operations Research,' " in *Science, Technology, and Warfare*, ed. Monte Wright and Lawrence Paszek (Washington, D.C.: Office of Air Force History, 1971), pp. 110–21.
25. Beard, *Developing the ICBM*, p. 70. See also pp. 211 and 217.
26. Elting Morison, *Men, Machines, and Modern Times* (Cambridge, Mass.: MIT Press, 1984), pp. 17–44, 98–122. I have also developed the concept of technological momentum in several essays, including "Technological Momentum," in *Does Technology Drive History?* ed. Merritt Roe Smith and Leo Marx (Cambridge, Mass.: MIT Press, 1994), pp. 101–13.
27. In a later essay about conservative momentum and change, Morison views matters differently. He empathizes with the naval establishment as it resisted, after the Civil War, at a time when sail power prevailed, the adoption of a steam-driven warship, the *Wampanoag*, even though it was excellently designed and performed more efficiently than sail-powered warships. He sympathizes with the naval officers who feared that a steam-driven ship would destroy essential naval values—a core culture—cultivated by the challenging experience of tending sailing ships. Morison accepts the argument of officers who believed that "acting as firemen and coal heavers will not produce in a seaman that combination of boldness, strength and skill which characterized the American sailor of an older day." He also condones the concern of officers who believed that commanders of steamships with their means of propulsion hidden belowdecks would not learn the habits of "observation, promptness and command found only on the deck of a sailing vessel." On the other hand, officers opposing the introduction of the steam-driven *Wampanoag* in peacetime did concede that the formidable ship might be a wartime necessity.
28. Schwiebert, "USAF's Ballistic Missiles," p. 68.
29. Beard, *Developing the ICBM*, p. 70, quoting Bush in *Hearings: Inquiry into Satellite and Missile Programs*, pp. 82–83.

30. Edmund Beard concludes that warhead weight was a false issue. Beginning in 1948–49, there were plans to develop nuclear warheads for the Snark and Navaho cruise missiles, evidence that warhead weight did not prevent long-range missiles. Beard, *Developing the ICBM*, p. 141.

31. Hanson Baldwin in the *New York Times*, 7 May 1950, and quoted in Beard, *Developing the ICBM*, p. 122.

32. Beard, *Developing the ICBM*, pp. 124–28.

33. Neufeld, *The Development of Ballistic Missiles*, p. 69.

34. Author's interview with General Bernard Schriever at his home in Washington, D.C., 19 November 1992. In contrast, J. Robert Oppenheimer, who headed the Manhattan Project and who had greater public prestige and influence among fellow scientists, alarmed the Air Force by his opposition to the near-term development of thermonuclear weapons.

35. Originally code-named Tea Garden for Trevor Gardner; its name was changed to Teapot after he found the other name had already been taken.

36. Author's interview with Simon Ramo at his office on Sunset Boulevard in Los Angeles, California, 13 June 1995.

37. Author's interview with Alain Enthoven in Palo Alto, California, 17 June 1995.

38. "Recommendations of the Tea Pot Committee," submitted to the committee by Simon Ramo, with an accompanying letter, on 1 February 1954, p. 2. Copies supplied to author by General Schriever.

39. According to Simon Ramo, the ICBM project remains "the biggest project we [U.S.] have ever had." The Apollo moon project might have been larger if it had not been able to use the industrial base laid down by the ICBM Project. Author's interview with Simon Ramo.

40. Beard, *Developing the ICBM*, p. 66.

41. Author's interview with Simon Ramo.

42. Beard, *Developing the ICBM*, p. 167.

43. Author's interview with General Bernard Schriever.

44. *New York Times*, 2 October 1963.

45. William Aspray, *John von Neumann and the Origins of Modern Computing* (Cambridge, Mass.: MIT Press, 1990), p. 91. On von Neumann, see also Steve Heims, *John von Neumann and Norbert Wiener* (Cambridge, Mass.: MIT Press, 1980).

46. Herbert York, *Making Weapons, Talking Peace: A Physicist's Odyssey from Hiroshima to Geneva* (New York: Basic Books, 1987), p. 89.

47. Author's interview with General Bernard Schriever.
48. York, *Making Weapons*, p. 64.
49. Author's interview with Simon Ramo.
50. Ibid.
51. "Recommendations of the Tea Pot Committee."
52. Ibid., p. 7.
53. Simon Ramo, "Background Facts on Convair in the ICBM Program," 24 April 1956, Archives, Ballistic Missile Organization (hereafter Archives BMO), p. 2.
54. Trevor Gardner, May 1956, quoted in Neufeld, *The Development of Ballistic Missiles*, p. 93.
55. Simon Ramo, "Background Facts on Convair," p. 2.
56. "Recommendations of the Tea Pot Committee," p. 9.
57. Committee member Bode, from AT&T's Bell Laboratories, and committee adviser Quarles, formerly of AT&T's Western Electric Company, as well as Ramo, had firsthand experience with such an approach.
58. Interview with Simon Ramo.
59. Simon Ramo, *The Business of Science: Winning and Losing in the High-Tech Age* (New York: Hill & Wang, 1988), p. 92.
60. Beard, *Developing the ICBM*, p. 172.
61. Ibid., pp. 166–67.
62. U.S. Air Force, Headquarters, Air Research and Development Command, Office of Information Services Historical Division, *The First Five Years of the Air Research and Development Command: United States Air Force* (Baltimore, 1955), pp. 17–24; and Beard, *Developing the ICBM*, pp. 108–11, 119.
63. On Schriever, see John Lonnquest, "The Face of Atlas: General Bernard Schriever and the Development of the Atlas Intercontinental Ballistic Missile, 1953–60" (Ph.D. diss., Duke University, 1996). For an overview, see, Claude J. Johns, "The United States Air Force Intercontinental Ballistic Missile Program, 1954–1959: Technological Change and Organizational Innovation" (Ph.D. diss., University of North Carolina, 1964).
64. "Man in the News," *New York Times*, December 1957, quoted in Neufeld, *The Development of Ballistic Missiles*, p. 108.
65. "Armed Forces: The Bird and the Watcher," *Time*, 1 April 1957, p. 18.
66. Author's interview with General Bernard Schriever.
67. Ibid. Lonnquest, "Face of Atlas," p. 53.

68. Author's interview with Charles Flagle, William Huggins, Robert Roy, and Alexander Kossiakoff at the Johns Hopkins Applied Physics Laboratory, 3 April 1995. Colonel Otto Glasser, who served on Schriever's WDD staff, found him a gifted planner and superb organizer, but not well organized in carrying out his daily routines.

69. Ramo, "Background Facts on Convair," p. 5.

70. Memorandum from B. A. Schriever to Major Robert T. Franzel, 4 August 1954, Archives BMO.

71. Minutes of Scientific Advisory Committee Meeting of 20–21 July 1954, Archives BMO.

72. Ramo, *The Business of Science*, pp. 35–36.

73. Memorandum from B. A. Schriever to Major Robert T. Franzel, 4 August 1954, Archives BMO. For an account of an airframe company moving to a systems or project management approach, see Glenn Bugos, "Manufacturing Certainty: Testing and Program Management for the F-4 Phantom II," *Social Studies of Science* 23 (1993): 265–300; and Glenn Bugos, *Engineering the F-4 Phantom II: Parts into Systems* (Annapolis, Md.: Naval Institute Press, 1996).

74. Author's interview with Francis Clauser at California Institute of Technology, 14 February 1992.

75. Walter Vincenti, a Stanford University aeronautical engineer, recalls that the Raytheon Company, an electronics firm, claimed it should do systems engineering for a fighter plane because it was little more than a platform for electronics. Interview with Vincenti at Stanford University, 8 July 1992.

76. Ramo, "Background Facts on Convair," p. 5.

77. "A Study of the Development Management Organization for the Atlas Program," 18 August 1954, Archives BMO. This document diagrams the "Convair Preferred Plan," the "Present Organizational Concept," and the plan using Ramo-Wooldridge as systems engineer and for technical direction.

78. Ramo-Wooldridge established its headquarters early in August 1954 in the chapel of a rented schoolhouse at 409 East Manchester Boulevard, Inglewood, California, near the present site of the Los Angeles airport. For security reasons, Schriever and his staff used the same building but were not identified as occupants.

79. Memorandum from B. A. Schriever to Major Robert T. Franzel, 4 August 1954, Archives BMO.

80. Ibid.

81. "A Study of the Development Management Organization for the Atlas Program."

82. B. A. Schriever to Lieutenant General T. S. Power, Commander Air Research and Development Command, 6 August 1954, Archives BMO.

83. Author's interview with General Bernard Schriever at his home in Washington, D.C., 29 December 1992.

84. "A Study of the Development Management Organization for the Atlas Program."

85. Alexander Kossiakoff, "The Systems Engineering Process," in *Operations Research and Systems Engineering,* ed. Charles D. Flagle, William H. Huggins, and Robert H. Roy (Baltimore: Johns Hopkins University Press, 1960), p. 102.

86. For a contrary argument, see Paul Forman, "Behind Quantum Electronics: National Security as the Basis for Physical Research in the United States, 1940–1960," *Historical Studies in the Physical and Biological Sciences* 18, no. 1 (1987): 149–229; and Stuart Leslie, *The Cold War and American Science* (New York: Columbia University Press, 1992).

87. Schriever to Lieutenant General Power, ARDC, 23 August 1954, Archives BMO; "A Study of the Development Management Organization for the Atlas Program."

88. "A Study of the Development Management Organization for the Atlas Program."

89. The civilian engineers and scientists had no "corner on brains." Author's interview with General Charles H. Terhune at his home in La Canada, California, 12 June 1995. The officers' rapid rise to positions of responsibility tends to support Terhune's evaluation: Colonel Otto J. Glasser, initially the warhead officer, became project officer for the Atlas missile; Colonel Benjamin Blasingame, initially responsible for guidance, became project officer for the sister Titan missile; and Colonel Edward Hall, initially in charge of propulsion, became project officer for the Thor missile. In March 1955, the list of telephone numbers of key WDD personnel includes thirteen colonels and lieutenant colonels and two majors. "Key WDD Names and Telephone Numbers," 30 March 1955, Archives BMO.

90. The expression is taken from Samuel C. Phillips, "Comments Concerning Management of Systems Acquisitions," in B. A. Schriever, "Report to the Packard Commission on the Defense Acquisition

Process," 1986, pp. 1, 3. Report given to author by General Schriever.

91. Schriever, "Report to the Packard Commission on the Defense Acquisition Process," pp. 7–8.

92. Author's interview with General Bernard Schriever, 29 December 1992.

93. Schriever, "Report to the Packard Commission on the Defense Acquisition Process," p. 1; Lonnquest, "Face of Atlas," p. 113.

94. Beard, *Developing the ICBM*, p. 190.

95. Author's interview with General Bernard Schriever, 29 December 1992.

96. Beard, *Developing the ICBM*, p. 188.

97. Office of the Assistant Secretary of the Air Force to Members of the Joint Evaluation Group, 15 September 1955, Archives BMO.

98. Secretary of Defense to Secretary of the Air Force, "Memorandum: Management of the ICBM and IRBM Development Programs," Office of the Secretary of Defense, Washington, D.C., 1955, Tab. A-5, copy supplied to author by General Bernard Schriever. The defense secretary's management policies applied to the Air Force's ICBM and IRBM programs and to the Army-Navy IRBM program.

99. "Memorandum: Management of the ICBM and IRBM Development Programs."

100. Ramo, *The Business of Science*, pp. 112–13.

101. Ibid., p. 113.

102. Schriever, "Report to the Packard Commission on the Defense Acquisition Process", Stephen D. Johnson, "From Concurrency to Phased Planning: An Episode in the History of Systems Management" (paper presented at the "Spread of the Systems Approach" conference held at the Dibner Institute, MIT, Cambridge, Mass., May 1996). Lonnquest presents concurrency as the essence of Schriever's management style and of his commitment to an embracing systems approach. "Face of Atlas," pp. 161ff.

103. Ramo, *The Business of Science*, p. 96.

104. B. A. Schriever to Colonel T. Harris, 3 August 1954, Archives, BMO.

105. Trevor Gardner, "How We Fell Behind in Guided Missiles," *Air Power Historian*, January 1956, p. 10.

106. Neufeld, *The Development of Ballistic Missiles*, pp. 119, 126; Simon Ramo, "Background Facts on Convair," p. 4.

107. Harvey Sapolsky, *The Polaris System Development* (Cambridge, Mass.: Harvard University Press, 1972); Graham Spinardi, *The Development of the U.S. FBM Technology: Polaris to Trident* (Cambridge: Cambridge University Press, 1993); R. A. Fuhrman, *Fleet Ballistic Missile System: Polaris to Trident* (Washington, D.C.: Lockheed Missiles & Space Company, 1978).

108. To compare Ramo with earlier leaders in aviation and aerospace, see Wayne Biddle, *Barons of the Sky* (New York: Simon & Schuster, 1991).

109. Ramo, *The Business of Science,* p. 32.

110. This biographical section on Simon Ramo is drawn mostly from "Entrepreneurs of the West: Simon Ramo, Interviewed by Christian G. Pease," Oral History Program, University of California, Los Angeles, 1985.

111. Ibid., p. 90.

112. Ibid., p. 118.

113. Ibid., p. 156.

114. *Fields and Waves in Modern Radio* has been revised periodically with the assistance of two coauthors in addition to Ramo. He denotes his royalties to universities.

115. Ramo, *The Business of Science,* p. 31.

116. "Entrepreneurs of the West," p. 245.

117. Author's interview with Simon Ramo, 13 June 1995.

118. "Entrepreneurs of the West," p. 225. Howard Hughes refused to be fingerprinted, so he could not obtain a security clearance that would allow him to learn about the classified work of Ramo's division.

119. "Entrepreneurs of the West," p. 266.

120. Ibid., p. 258.

121. Ibid., p. 276.

122. Raymond Puffer, "The History of ICBM Development," Air Force Systems Command, Norton Air Force Base, California, Archives BMO, p. 2.

123. Author's interview with Simon Ramo.

124. This definition is based on that given by Ramo. Author's interview with Simon Ramo.

125. Isaiah Berlin, *The Hedgehog and the Fox: An Essay on Tolstoy's View of History* (New York: Simon & Schuster, 1953), p. 1.

126. For more on testing, see Bugos, "Manufacturing Certainty," pp. 265–300.

127. Schriever to Commander of Air Research and Development Command, 28 January 1955, Archives BMO.

128. The process is detailed in Charles H. Terhune to Bernard Schriever, 8 December 1954; Otto J. Glasser to Terhune, 27 January 1955; and trip report of Otto Glasser, 25 January 1955, Archives BMO.

129. Charles Terhune, Jr., to Bernard Schriever, 8 December 1954, Archives BMO.

130. Otto J. Glasser, memorandum for the record, 25 January 1955, Archives BMO.

131. Otto J. Glasser to Charles Terhune, 27 January 1955, Archives BMO.

132. Otto J. Glasser, memorandum for the record, 7 February 1955; Glasser to Charles Terhune, 27 January 1955, Archives BMO.

133. Author's interview with Philip Caldwell in Pasadena, California, 12 June 1995. Caldwell, an engineer, was a member of the program office team. The "Black Saturday" meetings may have tapered off to monthly ones. Lonnquest, "Face of Atlas," p. 230.

134. Author's interview with General Bernard Schriever, 29 December 1992.

135. "Armed Forces: The Bird and the Watcher," Time, 1 April 1957, p. 19.

136. Author's interviews with General Bernard Schriever, 29 December 1992, and Colonel Terhune, 12 June 1995. Colonel Glasser dismissed the project control room as window dressing because the project managers already had the information. Schriever used it, according to Glasser, to impress visitors as representing advanced management. Lonnquest, "Face of Atlas," p. 228.

137. Chapman, Atlas, p. 95.

138. Lieutenant Colonel John Dodge to Colonel H. Norton, 6 April 1956, Archives BMO.

139. Historians of science have effectively demonstrated for their readers the role of experiment in the development of science; historians of technology have yet to convey to their readers the analogous importance of testing.

140. Lieutenant Colonel Lawrence Ely, WDD memorandum, June 1956, Archives BMO.

141. WDD memoranda for the record, 11 July 1956, 12 July 1956, and 16 July 1956, Archives BMO.

142. Louis G. Dunn of Ramo-Wooldridge Corporation to AF Ballistic Missile Division, 12 August 1957, Archives BMO.

143. Colonel Harold Norton to General Schriever, 11 October 1956, Archives BMO.

144. Lieutenant Colonel John Dodge, Memorandum: Comments on the Nose Cone Program, 4 December 1956, Archives BMO.

145. Louis G. Dunn to C. H. Terhune and J. A. Dodge, 8 March 1957, Archives BMO.

146. Colonel Dodge to Colonel Terhune, 15 January 1958; "Nose Cone Status," WDD memorandum, 26 June 1958; R. F. Mettler to WDD officers, 28 July 1958, Archives BMO.

147. Ramo, "Background Facts on Convair," p. 2.

148. F. R. Collbohm to John von Neumann, 14 October 1954, Archives BMO.

149. Bernard Schriever memorandum to Lieutenant General Power, 18 December 1954, Archives BMO.

150. Neufeld, *The Development of Ballistic Missiles*, p. 127.

151. Schriever memorandum to Lieutenant General Power, 24 February 1955.

152. Neufeld, *The Development of Ballistic Missiles*, p. 128.

153. Author's interview with General Bernard Schriever, 19 November 1992.

154. Ibid., 29 December 1992.

155. Ibid., 19 November 1992.

156. Beard, *Developing the ICBM*, pp. 150, 155, 182–83.

157. Author's interview with Gene Armstrong at his home in Westminster, California, 14 June 1995.

158. Ibid. Gene Armstrong, a design engineer and specialist in guidance and control at Convair in 1954, later became chief engineer for the astronautics division of General Dynamics, which had absorbed Convair. He later became chief engineer for the Defense Systems Group of TRW, which had absorbed Ramo-Wooldridge.

159. Not only Convair but also Ramo-Wooldridge received criticism from an Air Force officer who served as a principal on Schriever's WDD staff. In his book *The Art of Destructive Management: What Hath Man Wrought?* (New York: Vantage Press, 1984), Colonel Edward Hall, formerly in charge of the propulsion section of Atlas and project manager for the Thor missile project, starkly expressed his negative feelings about Ramo-Wooldridge. Interviewed after leaving the Air Force in 1960–61, Hall said flatly that incorporating a civilian organization, namely Ramo-Wooldridge, into the Air Force management system "was the biggest mistake ever made. . . . Schriever . . . regards that as his greatest accomplishment, I think it's done the Air Force a *tremendous* amount of damage." (Edward N.

Hall, interview with Jacob Neufeld [Washington, D.C.: Air Staff History Branch, Office of Air Force History, 1989], transcript, p. 1.) According to Hall, Ramo, after leaving the Hughes Company, placed "a half dozen kids" in a barbershop in Inglewood to establish the nucleus for the Ramo-Wooldridge systems engineering organization. Ramo, Hall recalls, hoped to sell equipment to the armed services, but when that did not "work out awfully well," Ramo asked von Neumann's permission to join the Teapot Committee, and "Johnny thought, 'Gosh, the guy's in the dumps, why not?'" Hall places this request and response in June 1954, which does not accord with the fact that Ramo had become a member of the committee when it was formed in November 1953. In general, Hall's account of how Ramo-Wooldridge became systems engineer and technical director differs from the one we have found in the records of WDD, especially the Schriever correspondence, and from my interviews with Schriever and Simon Ramo. Hall also recalls that Ramo used his influential position on the committee to argue that traditional Air Force contractors could not possibly perform a task as difficult as developing an ICBM and that the Atlas Project, therefore, needed scientific experts of the kind that Ramo-Wooldridge would employ. This was "absolute nonsense," Hall contends, because the problems of designing a space vehicle are less demanding than those of designing an airplane to operate in the atmosphere. "Celestial mechanics," he adds, "is a very old, well established art." The project did not need "great scientists," Hall adds, "it was strictly engineering." In Hall's sharp criticism, we note an underlying tension between scientists and engineers working together on research and development projects. Hall also laments the failure of Schriever to appreciate the important role that Air Force officers who had served, and were serving, at the Air Force's Wright-Patterson research and development center—such as Hall—ought to have had in the WDD. They could have advised effectively on the choice of contractors and monitored their work as well, Hall argues in the interview. Because Schriever's small staff of officers had, according to Hall, limited engineering experience, they could not appreciate his argument, choosing to rely on "these great scientists." (Hall, interview with Jacob Neufeld, pp. 2–6.) Hall sees the systems engineering organization as radically usurping the traditional responsibilities of engineers from aircraft companies and the military service. "All the deficiencies of short-sighted, dollar-oriented, civilian management, and some new ones, were to be imposed on WDD." "Cash flow, income, and salary levels" drove the Atlas Project, not a determination to define an engineering program incorporating "sufficient reach into the unknown to stimulate ingenuity, serendipity, and enthusiasm while eschewing the unattainable and pedestrian." (Hall, *The Art of Destructive Management*, p. 42.) The prominence of scientists in the Ramo-Wooldridge organization galled engineer Hall because they emphasized a scientific, or theoretical, approach as contrasted with an empirical and testing one. As the Atlas Project got under way, Hall, whose speciality was engine development, recalls that he had to "overcome the massive ignorance of Ramo-Wooldridge," whose scientists believed that they could design engines by using theory and mathematical analysis in the same way that they had designed electronic components. Hall, apparently, was not greatly impressed by Louis Dunn of the Ramo-Wooldridge staff; Dunn had been head of the Jet Propulsion Laboratory, which pioneered in the field of rocket propulsion.

Hall declares that "electrons obey linear differential equations and you can handle that mathematics quite easily. Unfortunately, reacting gases do not." (Hall, interview with Jacob Neufeld, p. 7.) Hall contends that he won this particular argument by relying on "massive empiricism" to develop the large liquid-fuel rocket engines, an approach made possible by the use of a new array of government-funded test facilities. (Hall, *The Art of Destructive Management*, pp. 40–41.) He further avows that he anticipated the design of the next-generation solid-fuel Minuteman missile but was not allowed to pursue his vision. He seems to have caused sufficient stir by his opposition to management policy and his single-minded campaign for a solid-fuel missile that he was relieved of his responsibilities as project manager for the Thor intermediate-range missile. (Beard, *Developing the ICBM*, p. 218 n.) Ramo, on the other hand, recalls that a number of his associates realized that solid-fuel was the next development. Schriever remembers Hall as a malcontent who had to be relieved as project manager. Caution, Schriever adds, was Hall's strong suit; he was good technically but highly emotional in his reaction to opposition, with little tolerance for seemingly intractable problems. (Author's interview with General Bernard Schriever, 29 December 1992.)

160. Author's interviews with Harry Charles, Ramo-Wooldridge test engineer and program manager, in July 1995. Other sources have

missile D as the first postdevelopment missile. Lonnquest, "Face of Atlas," p. 213.

161. Author's interviews with Harry Charles.

162. Ramo, *The Business of Science*, p. 100.

163. U.S. Congress, House of Representatives, Committee on Science and Astronautics, *A Chronology of Missile and Astronautic Events* (Washington, D.C.: Government Printing Office, 1961).

164. Author's interview with General Bernard Schriever, 29 December 1992.

165. Author's interview with Gene Armstrong, 14 June 1995.

166. Raymond Puffer, "The History of ICBM Development," Air Force Systems Command, Norton Air Force Base, California, Archives BMO, pp. 3, 4. Cost estimates for the development of the Atlas and Titan missiles vary, but one authority offers a figure of $2.5 billion. Deployment may have raised the figure to $17 billion. Charles J. Hitch, *Decision-Making for Defense* (Berkeley: University of California Press, 1965).

167. "Brief History of Aerospace Corporation," in U.S. Congress, House of Representatives, Committee on Government Operations, Military Operations Subcommittee, *Holifield Report*, 1962, p. 1129; William Leavitt, "Aerospace Corporation: USAF's Missile/Space Planning Partner," *Air Force/Space Digest*, October 1967, pp. 75ff. The founding of the Aerospace Corporation and the related establishment of Ramo-Wooldridge's Space Technology Laboratory followed upon congressional and other criticism of Ramo-Wooldridge's expanding role in the missile program. For a summary of this complex history, see Lonnquest, "Face of Atlas," pp. 260ff.

168. Ramo, *The Business of Science*, p. 100. For a sharp criticism of the role of Ramo Wooldridge in the ICBM project, see H. L. Nieburg, *In the Name of Science* (Chicago: Quadrangle Books, 1966).

169. Statement of Schriever on 1 August 1962, in *Holifield Report*.

170. Ramo, *The Business of Science*, p. 102.

IV. SPREAD OF THE SYSTEMS APPROACH

1. I am especially indebted to Fred Quivik, Atsushi Akera, and Erik Rau, research assistants on the Mellon Systems Project at the University of Pennsylvania, for research, essays, and insights helpful to me in composing this chapter. David Foster, formerly an undergraduate student at the University of Pennsylvania, also provided valuable research assistance in gathering materials for this chapter.

2. In 1964, the Systems Science Committee of the Institute of Electrical and Electronic Engineers, the world's largest professional society, defined "systems science" as a concept embracing operations research, systems analysis, and systems engineering. For an early discussion of the systems approach, see C. West Churchman, *The Systems Approach* (New York: Dell, 1968). For a systems and interdisciplinary approach to engineering problem solving, see Stephen J. Kline, *Conceptual Foundations for Multidisciplinary Thinking* (Stanford, Calif.: Stanford University Press, 1995).

3. The systems approach and the influence of experts spread abroad as well as in the United States. See Gabrielle Hecht, "Planning a Technological Nation: Systems Thinking and the Politics of National Identity in Postwar France"; Arne Kaijser and Joar Tiberg, "From Operations Research to Futures Studies: The Establishment, Diffusion, and Transformation of the Systems Approach in Sweden, 1945–1980" (papers presented at the "Spread of the Systems Approach" conference held at the Dibner Institute, MIT, Cambridge, Mass., May 1996).

4. A notable exception being the introduction by Merritt Roe Smith and the essays in *Military Enterprise and Technological Change: Perspectives on the American Experience* (Cambridge, Mass.: MIT Press, 1985). See also the discussion of the American systems of production in David A. Hounshell, *From the American System to Mass Production, 1800–1932: The Development of Manufacturing Technology in the United States* (Baltimore: Johns Hopkins University Press, 1984); and David W. Noble, *Forces of Production: A Social History of Industrial Automation* (New York: Knopf, 1984). For detailed military-history bibliographies emphasizing technology in the military, see Barton C Hacker, "Military Institutions, Weapons, and Social Change: Toward a New History of Military Technology," *Technology and Culture* 35 (October 1994): 768–834; and Alex Roland, "Science and War," in *Historical Writing on American Science,* ed. Sally Kohlstedt and Margaret Rossiter (Baltimore: Johns Hopkins University Press, 1986).

5. On military influences on science and engineering, see Paul Forman, "Behind Quantum Electronics: National Security as a Basis for Physical Research in the United States, 1940–1960," *Historical Studies in the Physical and Biological Sciences* 18, no. 1 (1987): 149–229; and Stuart Leslie, *The Cold War and American Science* (New York: Columbia University Press, 1992).

6. For the use of the systems concept during World War II, see David A. Mindell, "Automation's Finest Hour: Radar and System Integration in World War II" (paper presented at the "Spread of the Systems Approach" conference held at the Dibner Institute, MIT, Cambridge, Mass., May 1996).

7. Systems as a metaphor spread into fields only remotely related to engineering and management. Lily Kay, "How a Genetic Code Became an Information System" (paper presented at the "Spread of the Systems Approach" conference held at the Dibner Institute, MIT, Cambridge, Mass., May 1996); Evelyn Fox Keller, *Refiguring Life: Metaphors of Twentieth-Century Biology* (New York: Columbia University Press, 1995).

8. Author's interview with Alexander Kossiakoff at Johns Hopkins University Applied Physics Laboratory, 3 April 1995. For recent views on complex systems, see Lynn Nadel and Daniel Stein, eds., *Lectures in Complex Systems: The Proceedings of the 1991 Complex Systems Summer School, Santa Fe, New Mexico, June 1991* (Reading, Mass.: Addison-Wesley, 1992).

9. John N. Warfield and J. Douglas Hill, *A Unified Systems Engineering Concept*, Battelle Monographs, ed. Benjamin Gordon (Columbus, Ohio: Battelle, 1972), p. 1.

10. Quoted in Melvin S. Day, "Space Technology for Non-Aerospace Applications," *1968 Wescon Technical Papers*, vol. 12, pt. 5, session 23, paper 4 (1968), pp. 5–6.

11. Warren G. Bender, "A Survey of Systems Science Education," *IEEE Transactions on Systems Science and Cybernetics*, vol. SSC-1 (November 1965), pp. 26–31 (see table). Albert Rubenstein, "A Program of Research on the Research and Development Process," *IEEE Transactions on Engineering Management*, September 1964, p. 113.

12. Marvin Berkowitz, *The Conversion of Military-Oriented Research and Development to Civilian Uses* (New York: Praeger, 1970), pp. 255ff. Berkowitz drew on John S. Gilmore, John J. Ryan, and William S. Gould, *Defense Systems Resources in the Civil Sector: An Evolving Approach, An Uncertain Market*, prepared for the U.S. Arms Control and Disarmament Agency by the Denver Research Institute (Washington, D.C.: Government Printing Office, 1967), to make these estimates.

13. Alexander Kossiakoff, "The Systems Engineering Process," in *Operations Research and Systems Engineering*, ed. Charles D. Flagle, William H. Huggins, and Robert H. Roy (Baltimore: Johns Hopkins University Press, 1960), pp. 82–118.

14. Ibid., p. 102.
15. C. West Churchman, Russell L. Ackoff, and E. Leonard Arnoff, *Introduction to Operations Research* (New York: Wiley, 1957), p. 1. In this section on operations research, I am drawing on two unpublished essays, "Operations Research" and "Common Good to Common Defense: Political Migrations and Institutionalization of Operations Research, 1937–ca. 1961" by Erik Peter Rau, which he prepared as a graduate student and research assistant on this book project.
16. W. E. Duckworth, *A Guide to Operational Research* (London: Methuen, 1965), p. 8.
17. Philip McCord Morse and George Kimball, *Methods of Operations Research* (Cambridge, Mass.: Technology Press of MIT, 1951), p. 1.
18. Solly Zuckerman, *From Apes to Warlords: The Autobiography (1904–1946) of Solly Zuckerman* (London: Hamish Hamilton, 1978), p. 266.
19. J. D. Bernal, *Science in History*, 4 vols. (1954; reprint, Cambridge, Mass.: MIT Press, 1971), 3:833.
20. Robert H. Roy, "The Development and Future of Operations Research and Systems Engineering," in *Operations Research and Systems Engineering,* ed. Charles D. Flagle, William H. Huggins, and Robert H. Roy (Baltimore: Johns Hopkins University Press, 1960), p. 9.
21. Stephen P. Waring, *Taylorism Transformed: Scientific Management Theory Since 1945* (Chapel Hill: University of North Carolina Press, 1991), p. 21.
22. Rau, "Operations Research," pp. 4–5. See also M. Fortun and S. S. Schweber, "Scientists and the Legacy of World War II: The Case of Operations Research," *Social Studies of Science,* 23 November 1991, pp. 595–642.
23. Erik Rau, "The Adoption of Operations Research in the United States During World War II" (paper presented at the "Spread of the Systems Approach" conference held at the Dibner Institute, MIT, Cambridge, Mass., May 1996).
24. I. B. Holley, "The Evolution of Operations Research and Its Impact on the Military Establishment: The Air Force Experience," in *Science, Technology, and Warfare,* ed. Monte Wright and Lawrence Paszek (Washington, D.C.: Office of Air Force History, 1971), pp. 90–91.
25. Rau, "Operations Research," p. 10.

NOTES TO PAGES 151-160

26. Philip M. Morse, "Operations Research, What is It?" in *Proceedings of the First Seminar in Operations Research, November 8–10, 1951* (Cleveland, Ohio: Case Institute of Technology, 1951): p. 39.

27. Philip M. Morse, *In at the Beginnings: A Physicist's Life* (Cambridge, Mass.: MIT Press, 1977), p. 291.

28. Ibid., p. 290.

29. Author's interview with Robert Roy at Johns Hopkins University Applied Physics Laboratory, 3 April 1995.

30. Author's interview with Charles Flagle at Johns Hopkins University Applied Physics Laboratory, 3 April 1995.

31. I am indebted to Fred Quivik of the University of Pennsylvania for this tabulation, which I have taken from one of a series of unpublished essays, "The Context for Systems Engineering in the 1960s," that he prepared in 1992 as a graduate student and research assistant on this book project.

32. Bruce L. R. Smith, *The RAND Corporation* (Cambridge, Mass.: Harvard University Press, 1966), p. 30.

33. Author's interview with Cullen Craig, Merton Davies, and Bruno Augustine at RAND headquarters in Santa Monica, California, 14 February 1992. William Leavitt, "RAND: The Air Force's Original Think Tank," *Air Force/Space Digest,* May 1967, pp. 100–106, 109.

34. Leavitt, "RAND," p. 104.

35. Daniel Bell, ed., *The Radical Right* (New York: Doubleday/Anchor, 1964), p. 33.

36. A. J. Wohlstetter et al., *Selection and Use of Strategic Air Bases,* Report R-266 (RAND Corporation, 1954).

37. Author's interview with Alain Enthoven in Palo Alto, California, 17 June 1995.

38. David Hounshell, "The Medium Is the Message, or How Context Matters: The RAND Corporation Builds an Economics of Innovation, 1946–1962" (paper presented at the "Spread of the Systems Approach" conference held at the Dibner Institute, MIT, Cambridge, Mass., May 1996).

39. Author's interview with Cullen Craig, Merton Davies, and Bruno Augustine.

40. Author's interview with Alain Enthoven.

41. Ibid.

42. Systems analysis and the use of computers spread into some government agencies before the Johnson order. Atsushi Akera, "Engineers or Managers? The Systems Analysis of Electronic Data Processing

in the Federal Bureaucracy" (paper presented at the "Spread of the Systems Approach" conference held at the Dibner Institute, MIT, Cambridge, Mass., May 1996).

43. Deborah Shapley, *Promise and Power: The Life and Times of Robert McNamara* (Boston: Little, Brown, 1993), p. 48.

44. Ibid., pp. 100–101.

45. Alain Enthoven and K. Wayne Smith, *How Much Is Enough? Shaping the Defense Program, 1961–1969* (New York: Harper & Row, 1971).

46. David Jardini, "Out of the Blue Yonder: The Transfer of Systems Thinking from the Pentagon to the Great Society, 1961–1965" (paper presented at the "Spread of the Systems Approach" conference held at the Dibner Institute, MIT, Cambridge, Mass., May 1996).

47. Fred Kaplan, *The Wizards of Armageddon* (New York: Simon & Schuster, 1983), p. 253.

48. Samuel A. Tucker, ed., *A Modern Design for Defense Decision: A McNamara-Hitch-Enthoven Anthology* (Washington, D.C.: Industrial College of the Armed Forces, 1966), pp. 126–27.

49. Kaplan, *The Wizards of Armageddon,* p. 254.

50. Ibid., pp. 121, 244, 254.

51. Alain C. Enthoven, "Choosing Strategies and Selecting Weapon Systems," in Tucker, *A Modern Design for Defense Decision,* p. 137.

52. Ibid., p. 141.

53. Author's interview with Alain Enthoven.

54. Ibid.

55. Daniel Seligman, "McNamara's Management Revolution," *Fortune,* July 1965, pp. 117ff.

56. Author's interview with Alain Enthoven.

57. Ibid.

58. In 1970, Marvin Berkowitz observed that there were three catch-phrases in vogue in government circles: "systems analysis," "quality of life," and "social indicators." Berkowitz, *The Conversion of Military-Oriented Research and Development to Civilian Uses,* pp. 244–45.

59. John N. Warfield, *An Assault on Complexity* (Columbus, Ohio: Battelle, 1973), p.i.

60. On crises of control, see James R. Beniger, *The Control Revolution: Technological and Economic Origins of the Information Society* (Cambridge, Mass.: Harvard University Press, 1986).

61. Zygmunt Bauman, *Intimations of Postmodernity* (London: Routledge, 1992); David Harvey, *The Condition of Postmodernity: An Enquiry into the Origins of Cultural Change* (Oxford: Basil Blackwell, 1989).

62. For a bibliography of works on complexity see David Warsh, *The Idea of Complexity* (New York: Viking, 1984), pp. 3–4, 198–211. Recently, Murray Gell-Mann has explored complexity in *The Quark and the Jaguar: Adventures in the Simple and the Complex* (New York: W. H. Freeman, 1995). Joel Moses, mathematician, computer scientist, and provost at the Massachusetts Institute of Technology, has defined a complex system succinctly as a system with many nonsimple, non-pattern-repeating connections. He does not equate complexity with the number of functions the system is able to perform. The heterogeneity—as contrasted with the homogeneity—of the components of a system contribute to its complexity. His definition is abstract enough to cover both technical and organic systems. Joel Moses, "Organization and Ideology: Paradigm Shifts in Systems and Design," unpublished monograph, July 1991, pp. 114–21.

63. Robert Venturi, *Complexity and Contradiction in Architecture* (New York: Museum of Modern Art, 1966), p. 16.

64. Guy Black, "Systems Analysis in Government Operations," *Management Science*, 14 October 1967, p. B41.

65. As quoted in Jay W. Forrester, *Urban Dynamics* (Cambridge, Mass.: MIT Press, 1969), p. vii.

66. Simon Ramo, "The Systems Approach to Automated Common Sense," *Nation's Cities*, 6 March 1968.

67. Black, "Systems Analysis in Government Operations," p. B42.

68. Guy Black, *The Application of Systems Analysis to Government Operations* (New York: Praeger, 1968), pp. 1–3.

69. Robert H. Haveman, *Poverty Policy and Poverty Research: The Great Society and the Social Sciences* (Madison: University of Wisconsin Press, 1987), pp. 34, 161.

70. This section on TRW's civil programs draws on Davis Dyer, "The Limits of Technology Transfer: Civil Systems at TRW, 1965–1975" (paper presented at the "Spread of the Systems Approach" conference held at the Dibner Institute, MIT, Cambridge, Mass., May 1996).

71. Ramo's books include *Cure for Chaos* (1969), *Century of Mismatch* (1970), and *The Business of Science: Winning and Losing in the High-Tech Age* (1988).

72. "Cities and Systems Management," editorial, *Aviation Week & Space Technology*, 1 July 1968, 11.

73. "Urban Incorporated," a prospectus describing the purpose and structure of the consortium. I am grateful to General Bernard

Schriever for sharing these and other documents pertaining to Urban Incorporated and Urban Systems Associates.

74. Urban Systems Associates, "Proposal for a Study of Methods to Increase Industry's Involvement in the Improvement of Urban Conditions," c. 1968, p. 1

75. Ibid., p. 3.

76. Urban Systems Associates, "Final Report: The Idea, the Experience, and the Conclusions," 1969, p. 1.

77. Ibid., p. 13.

78. "Cities and Systems Management," editorial, *Aviation Week & Space Technology*.

79. Urban Systems Associates, "Final Report," p. 43.

80. Author's interview with Jay Forrester at MIT, 20 May 1994; Jay W. Forrester, "From the Ranch to System Dynamics," in *Management Laureates: A Collection of Autobiographical Essays,* ed. Arthur G. Bedeian (Greenwich, Conn.: JAI Press, 1992).

81. He had been a member of the MIT research staff assigned to a research center. This reminds us that a substantial share of MIT's creativity is found among persons with research appointments, not to be confused with faculty appointments.

82. Forrester, "From the Ranch to System Dynamics," p. 344.

83. Communication from Forrester to the author, 28 July 1997.

84. Jay W. Forrester, "System Dynamics—Future Opportunities," in *System Dynamics,* ed. Augusto A. Legasto, Jr., Forrester, and James M. Lyneis (Amsterdam: North-Holland, 1980), p. 13.

85. Forrester, "From the Ranch to System Dynamics," pp. 344–45.

86. Paul N. Edwards, "The World in a Machine: Origins and Impacts of Early Computerized Global Systems Models," pp. 27–28 (paper presented at the "Spread of the Systems Approach" conference held at the Dibner Institute, MIT, Cambridge, Mass., May 1996).

87. Forrester, *Urban Dynamics,* p. ix.

88. Forrester, "From the Ranch to System Dynamics," p. 348.

89. Ibid.

90. Forrester, *Urban Dynamics,* pp. 14–15.

91. Forrester, "System Dynamics—Future Opportunities," p. 16.

92. For a reaction of decision makers to Forrester's models, see Forrester, "From the Ranch to System Dynamics," pp. 348–49.

93. Forrester, *Urban Dynamics,* p. 10.

94. Forrester, "System Dynamics—Future Opportunities," p. 12.

95. Forrester, "From the Ranch to System Dynamics," p. 346.

96. Ibid., p. 348.
97. D. H. Meadows et al., *The Limits of Growth* (New York: Universe, 1972); Edwards, "The World in a Machine."
98. Carl Kaysen, "The Computer that Cried W*O*L*F*," *Foreign Affairs* 50 (1972): 660–68; Jay W. Forrester, *World Dynamics* (Portland, Ore.: Productivity Press, 1971).
99. Forrester, "System Dynamics—Future Opportunities," p. 14.
100. Daniel H. Elliott, chairman of the New York City Planning Commission, quoted in "Cities Still Grappling with Technology," *Aviation Week & Space Technology*, 7 August 1972, p. 65.
101. Ida R. Hoos, *Systems Analysis in Social Policy: A Critical Review* (London: Institute of Economic Affairs, 1969), p. 34.
102. "Space Coming Down to Earth," *The Economist*, 20 March 1965, pp. 1, 275.
103. Ida R. Hoos, *Systems Analysis in Public Policy* (Berkeley: University of California Press, 1972), p. 5.
104. Berkowitz, *The Conversion of Military-Oriented Research and Development to Civilian Uses*, pp. 597–604.
105. Dean S. Warren, "Human vs. Hardware—A Critical Look at Aerospace as an Urban Problem Solver," *Aviation Week & Space Technology*, 7 June 1971, p. 63. I am drawing, in this section on aerospace and urban problems, upon the unpublished collection of essays by Fred Quivik of the University of Pennsylvania entitled "The Context for Systems Engineering in the 1960s," which he prepared in 1992 as a graduate student and research assistant on this book project.
106. Warren, "Human vs. Hardware," p. 63.
107. "Aerospace's Urban Role Debated," *Aviation Week & Space Technology*, 7 June 1971, pp. 63–64.
108. Daniel P. Moynihan, "The Schism in Black America," *Public Interest*, Spring 1972, quoted in John N. Warfield, *An Assault on Complexity* (Columbus, Ohio: Battelle, 1973), p. 1-1. Jay Forrester now contends that he and Emeritus Dean Gordon Brown of MIT have responded to the public's difficulty in dealing with complex systems by introducing the study of system dynamics at the level of K–12 education. See his website at http://sysdyn.mit.edu.
109. Stewart Brand et al., *Whole Earth Catalog* (Menlo Park, Calif.: Portola Institute, 1968).
110. Theodore Roszak, *The Making of a Counter-Culture: Reflections on the Technocratic Society and Its Youthful Opposition* (Garden City, N.Y.: Doubleday, 1969).

III. John McDermott, "Technology: The Opiate of the Intellectuals," *New York Review of Books,* 31 July 1969, pp. 25–35.

112. Charles Reich, *The Greening of America* (New York: Bantam, 1971), pp. 92–93.

113. For more on the reaction against technological systems, see Thomas P. Hughes, *American Genesis: A Century of Invention and Technological Enthusiasm, 1870–1970* (New York: Viking, 1989), pp. 443–72.

114. Tom Peters, letter to the editor of the [Stanford] *Reporter,* 18 May 1995.

115. Robert S. McNamara, *In Retrospect: The Tragedy and Lessons of Vietnam* (New York: Times Books, 1995).

116. Author's interview with Alain Enthoven.

117. Author's interview with Russell Ackoff in Philadelphia, 4 November 1993.

118. Several Ackoff articles sum up his criticisms. "The Future of Operations Research Is Past," *Journal of the Operational Research Society* 30, no. 2 (1979): 93–104; "Resurrecting the Future of Operational Research," *Journal of the Operational Research Society* 30, no. 3 (1979): 189–99. For the earlier views of Ackoff, see "The Development of Operations Research as a Science," *Operations Research: The Journal of the Operations Research Society of America* 4, no. 3 (1956): 265–95.

119. Ackoff, "The Future of Operations Research Is Past," p. 93.

120. Author's interview with Russell Ackoff in Philadelphia, 4 November 1993. See Russell L. Ackoff, "The Art and Science of Mess Management," *Interfaces* 17 (1981): 20–26.

121. Ackoff, "The Future of Operations Research Is Past," p. 99.

122. Hoos, *Systems Analysis in Social Policy,* pp. 54–55.

123. Ibid., p. 33.

124. Ibid., pp. 36–37.

125. Ibid., p. 47.

126. For a balanced study of the role of experts, see Brian Balogh, *Chain Reaction: Expert Debate and Public Participation in American Commercial Nuclear Power, 1945–1975* (New York: Cambridge University Press, 1991).

V. COPING WITH COMPLEXITY: CENTRAL ARTERY/TUNNEL

1. The Berlin Project is located on a large site ravaged by World War II bombs and subsequent destruction to make way for the infamous Berlin Wall between the former East Berlin and West Berlin.

Internationally renowned architects have designed the public, commercial, recreational, and residential buildings, and the German National Railway is constructing a rail tunnel under the site connecting a new railway station in the north with rail lines coming into Berlin from the south. An automobile highway will parallel the rail tunnel. *Info Box: Der Katalog* (Berlin: Verlag Dirk Nishen, 1996).

2. Bruce E. Seely, *Building the American Highway System: Engineers as Policy Makers* (Philadelphia: Temple University Press, 1987). I am indebted to Richard John, director, Transportation Systems Center, Thomas Smith, Central Artery Project administrator, and Peter Markle, division administrator, at the U.S. Department of Transportation, Federal Highway Administration, Massachusetts Division, for an interview concerning various aspects of their division's role in CA/T (3 October 1997).

3. Antonio di Mambro, "Boston's Big Dig," *Spazio e Società (Space and Society)* 14 (April–June 1991), English preface and p. 35.

4. Former Massachusetts State Senate President Kevin Harrington, quoted in David Luberoff, Alan Altshuler, and Christie Baxter, *Mega-Project: A Political History of Boston's Multibillion Dollar Artery/Tunnel Project* (Cambridge, Mass.: Taubman Center, John F. Kennedy School, Harvard University, 1993), p. 10 (hereafter *Mega-Project*).

5. *Mega-Project*, p. 8.

6. Author's interview with Fred Salvucci at MIT, 10 February 1994.

7. Ibid.

8. Alan Lupo, *Rites of Way: The Politics of Transportation in Boston and the U.S. City* (Boston: Little, Brown, 1971), pp. 96–97, quoted in *Mega-Project*, p. 12.

9. Author's interview with John Wofford in Boston, 24 May 1994.

10. *Mega-Project*, p. 16.

11. Ibid., p. 21.

12. Author's interview with Fred Salvucci, 10 February 1994.

13. Ibid.

14. Kevin Harrington, former president of the State Senate, quoted in *Mega-Project*, p. 20.

15. Ibid., p. 28.

16. Ibid., p. 24.

17. Author's interview with Fred Salvucci, 10 February 1994.

18. *Mega-Project*, p. 86.

19. Ibid., p. 90.

20. Ibid., p. 97.

21. *Washington Post,* 28 March, 1987, quoted in *Mega-Project,* p. 125.

22. *Mega-Project,* p. 87.

23. Ibid., p. 111.

24. Ibid., p. 125.

25. Ibid., p. 129.

26. Ibid., pp. 83–84.

27. Ibid., pp. 135–36.

28. Ibid., p. 139.

29. Commonwealth of Massachusetts Department of Public Works, "Contract for State Highway Work," Parsons, Brinckerhoff, Quade Douglas–Bechtel Civil, Inc., 28 August 1986.

30. Commonwealth of Massachusetts Highway Department, "Contract for State Highway Work," 22 September 1993.

31. On joint ventures see Glenn Bugos, "System Reshapes the Corporation: Joint Ventures in the Bay Area Rapid Transit System, 1962–1972" (paper presented at the "Spread of the Systems Approach" conference held at the Dibner Institute, MIT, Cambridge, Mass., May 1996); David R. Dibner, *Joint Ventures for Architects and Engineers* (New York: McGraw-Hill, 1972).

32. Benson Bobrick, *Parsons Brinckerhoff: The First Hundred Years* (New York: Van Nostrand Reinhold, 1988).

33. Robert L. Ingram, *The Bechtel Story: Seventy Years of Accomplishment in Engineering and Construction* (San Francisco, 1968).

34. Robert Thomas Crow, "The Business Economist at Work: The Bechtel Group," *Business Economics,* 29 January 1994, p. 46.

35. Thomas P. Hughes, *Networks of Power: Electrification in Western Society, 1880–1930* (Baltimore: Johns Hopkins University Press, 1993), pp. 386–403.

36. David Weir, "Friends in High Places: The Bechtel Story: The Most Secret Corporation and How it Engineered the World" book review, *The Nation,* 17 April 1988, pp. 613–16; Laton McCartney, *Friends in High Places: The Bechtel Story* (New York: Simon & Schuster, 1988).

37. Mark Dowie, "The Bechtel File: How the Master Builders Protect Their Beachheads," *Mother Jones,* September–October 1978, pp. 29–38, reprinted in *1978 Investigative Reports* (San Francisco: Center for Investigative Reporting).

38. Weir, "Friends in High Places," pp. 613–16.

39. Jacob Weisberg, "The Boycott Memo: Bechtel Won't Touch Israel," *New Republic,* 8 September 1986, pp. 12–14.

40. Massachusetts Department of Public Works, *Central Artery (I-93)/Third Harbor Tunnel (I-90) Project, Boston, Massachusetts: Final Supplemental Environmental Impact Statement/Report, EOEA#4325* (November 1990).

41. Quoted in Thomas C. Palmer, Jr., "Commitments to Foes Raise Artery Price Tag," *Boston Globe,* 13 September 1994.

42. Author's interview with Fred Salvucci, 25 June 1994.

43. *Mega-Project,* pp. 207–11.

44. Palmer, "Commitments."

45. *Mega-Project,* p. 216.

46. K. Dun Gifford, "The Artery Project: Unnecessary Surgery," *Boston Globe,* 21 October 1990.

47. Ibid.

48. Author's interview with Peter Howe in Cambridge, Mass., 26 May 1994.

49. Peter Howe, "Environmentalists, Others Make Final Attempt to Halt Interchange," *Boston Globe,* 18 November 1990. Salvucci contends that the artist's rendition distorts the image by emphasizing verticality.

50. Peter Howe, "Red Flags on Scheme Z Rise at City Hall," *Boston Globe,* 4 December 1990. See also Howe *Globe* articles on December 2, 4, 7, 9, 13, 14, 20, 21, and 28.

51. Reporters, Howe feels, ought to arouse informed public skepticism about huge expenditures that affect the environment and community life. Author's interview with Peter Howe.

52. Peter Howe, "It'll Get You Across the Charles River, but at What Price? Critics Ask," *Boston Globe,* 9 December 1990.

53. Ibid.

54. Peter Howe, "Artery Scheme Z Scored As 'Nightmare' at Hearing," *Boston Globe,* 14 December 1990.

55. Author's interview with Fred Salvucci, 25 May 1964.

56. Peter Howe, "Will Artery Suffer Fate of NY's Westway?" *Boston Globe,* 27 January 1991.

57. Howe, "It'll Get You Across the Charles River."

58. Weinberg, quoted in *Mega-Project,* p. 226.

59. *Mega-Project,* p. 157.

60. Author's interview with Fred Salvucci, 25 May 1994.

61. Author's interview with Fred Salvucci, 10 February 1994.

62. Peter Howe, "Anti-Tunnel Group's Link Raises a Question," *Boston Globe,* 4 February 1991.

63. Peter Howe, "New Coalition Backs the Artery Project," *Boston Globe,* 21 December 1990.

64. Peter Howe, "Artery Scheme Z Scored As 'Nightmare' at Hearing."

65. Peter Howe, "DeVillars Gives Major OK to Artery Project," *Boston Globe,* 3 January 1991.

66. Ibid.

67. At the start of deliberations, Wofford asked each committee member to state her or his primary concerns and objectives; then he used these as a set of references against which to evaluate various modifications and alternatives for Scheme Z.

68. Author's interview with John Wofford, 24 May 1994; John G. Wofford, "Charles River Crossing: Bridge over Troubled Waters— An Update," 20 April 1994, a typescript supplied by Wofford.

69. Bridge Design Review Committee, "Report on the Charles River Crossing: Executive Summary," October 1991.

70. Wofford, "Charles River Crossing," p. 1.

71. Author's interview with Fred Salvucci, 25 May 1994.

72. Ibid.

73. Bridge Design Review Committee, "Report on the Charles River," p. 10.

74. Author's interview with Fred Salvucci, 25 May 1994.

75. I have not troubled the reader with the complicated titles given to environmental impact statements as they move through drafts and revisions. For example, see the draft for comment of the new EIS for the Charles River Crossing: Massachusetts Highway Department, *Charles River Crossing Central Artery/Tunnel Project, Boston, Massachusetts: Draft Supplemental Environmental Impact Statement/Report* (DSEIS/R). "EIS" refers to a document prepared at the behest of the Federal Highway Administration in response to federal law; "EIS/R" refers to a document responding to Massachusett's environmental law.

76. Thomas G. Palmer, Jr., "Decision on River-Crossing Design Sinks In," *Boston Globe,* 21 November 1993.

77. Ibid.

78. Quoted in Thomas G. Palmer, Jr., "All-Bridge Plan Cheapest Way to Go, Kerasiotes Says," *Boston Globe,* 17 November 1993.

79. Quoted, ibid.

80. Palmer, "Decision on River-Crossing Design Sinks In."

81. Author's interview with Tad Weigle and Jerry Riggsbee at CA/T headquarters, South Station, Boston, 28 October 1994.

82. Author's interview with Peter Zuk at CA/T headquarters, South Station, Boston, 4 February 1997.

83. Author's interview with General William S. Flynn at CA/T headquarters, South Station, Boston, 4 February 1997.

84. Ibid.

85. Author's interview with Jeff Brunetti at CA/T headquarters, South Station, Boston, 28 October 1993.

86. Author's interview with Tony Lancellotti at CA/T headquarters, South Station, Boston, 27 October 1993.

87. Author's interview with Bill Edwards at CA/T headquarters, South Station, Boston, 24 May 1994. I am indebted to Edwards for this account of the control center, which he directed along with Bob Marshall.

88. O. Ray Pardo, "Advanced Highway Control on the Central Artery," unpublished essay, 1993.

89. Author's interview with O. Ray Pardo at 185 Kneeland Street, Boston, 28 October 1993.

90. Author's interview with Hubert Murray at Wallace, Floyd Associates, 20 May 1994.

91. Bechtel/Parsons Brinckerhoff (B/PB), "Central Artery/Tunnel Project: Subconsultant Agreement No. 96159-012, Wallace, Floyd Associates, Inc., Architectural and Urban Design, Community Liaison and Construction Mitigation Services," 1 July 1996, Attachment D, p. 1.

92. Bechtel/Parsons Brinckerhoff, "Central Artery/Tunnel Project: Subconsultant Agreement No. 94165-012: Wallace, Floyd Associates, Inc.," 1 October 1993: Attachment D, p. 8

93. Author's interview with Hubert Murray, 20 May 1994.

94. Otile McManus, "Putting the Art in Artery," Boston Globe, 14 November 1990.

95. Di Mambro, "Boston's Big Dig," p. 30.

96. Thomas C. Palmer, Jr., "Mismanagement Blamed for Costs, Delays in Big Dig," Boston Globe, 12 December 1995.

97. Thomas C. Palmer, Jr., "Big Dig Gets Its Report Card: Consultant Outlines Five Areas of Concern," Boston Globe, 26 December 1995.

98. Peterson Consulting, Ltd., "Management Review Report," 1995, phase 1, p. 2.

99. Author's interview with Peter Zuk, 4 February 1997.

100. Peterson Consulting, Ltd., "Management Review Report," pp. 1–8.

101. Thomas C. Palmer, Jr., "Artery Project Official Ousted," Boston Globe, 22 November 1994.

102. Peterson Consulting, Ltd., "Management Review Report," p. 1-1.

VI. NETWORKING: ARPANET

1. Interview with Leonard Kleinrock conducted by Judy O'Neill on 3 April 1990, Charles Babbage Institute, Center for the History of Information Processing, 1990 (hereafter cited as Charles Babbage Institute). The Charles Babbage Institute interviews used in this chapter were conducted by staff members. The original transcripts are located in the center's archives at the University of Minnesota, Minneapolis. I use digital copies of the interviews.

2. The following author's interviews with engineers who were members of small teams working on large projects, such as the Atlas and the Navy's Polaris missile projects, confirm this supposition. Gene Armstrong at his home in Westminster, California, 14 June 1995; Robert Fuhrman in Washington, D.C., 27 June 1995; telephone interviews with Harry Charles, July 1995; and John Kimball, Bernie Pfefer, William Whitmore, and Howard Burnett in Mountain View, California, 19 June 1995.

3. Systems built upon early systems have been called "second-order systems." See Ingo Braun, *Geflügelte Saurier: Systeme zweiter Ordnung: eine Verflechtungsphänomen grosser technischer Systeme,* monograph from Wissenschaftszentrum Berlin für Sozialforschung, Berlin, n.d.

4. Paul Edwards, *The Closed World: Computers and the Politics of Discourse in Cold War America* (Cambridge, Mass.: MIT Press, 1996), pp. 210–13.

5. Interview with J. C. R. Licklider conducted by William Aspray and Arthur Norberg, 28 October 1988, Charles Babbage Institute.

6. Alexander A. McKenzie and David C. Walden, "ARPANET, The Defense Data Network and Internet," in *The Froelich/Kent Encyclopedia of Telecommunications,* ed. Fritz E. Froehlich (New York: Marcel Dekker, 1991), 1:344.

7. J. C. R. Licklider, "Man-Computer Symbiosis," *IRE Transactions on Human Factors in Engineering* 1, no. 1 (March 1960): 5.

8. *Webster's Second New International Dictionary* (Springfield, Mass.: Merriam, 1958).

9. Licklider, "Man-Computer Symbiosis," p. 5.

10. Ibid., p. 7.

11. For the continuity among projects, see Hans Dieter Hellige, "From Sage via Arpanet to Ethernet: Stages in Computer Communications Concepts Between 1950 and 1980," *History and Technology* 11 (1994): 49–75.

12. On the MIT electrical engineering department and its allied research centers, see Karl L. Wildes and Nilo A. Lindgren, *A Century of Electrical Engineering and Computer Science at MIT, 1882–1982* (Cambridge, Mass.: MIT Press, 1985), pt. 5.

13. The Licklider quotations in this section are from the interview with Licklider conducted by Aspray and Norberg, 28 October 1988.

14. On the evolution of electric light and power systems, see Thomas P. Hughes, *Networks of Power: Electrification in Western Society, 1880–1930* (Baltimore: Johns Hopkins University Press, 1983).

15. Fano quoted in Arthur Norberg and Judy E. O'Neill, *A History of the Information Processing Techniques Office of the Defense Advanced Research Projects Agency* (Minneapolis: Charles Babbage Institute, 1992), I:44.

16. Robert M. Fano, "Project MAC," in *Encyclopedia of Computer Science and Technology*, ed. Jack Belzer, Albert G. Holzman, and Allen Kent (New York: Marcel Dekker, 1979), 12:339–60. On the background of Project MAC, see J. McCarthy, "Time-Sharing Computer Systems," in *Management and the Computer of the Future*, ed. M. Greenberger (Cambridge, Mass.: MIT Press, 1962), pp. 221–36.

17. McCarthy quoted in Fano, "Project MAC," p. 343.

18. Martin Campbell-Kelly and William Aspray, *Computer: A History of the Information Machine* (New York: Basic Books, 1996), p. 218; Jeremy Main, "Computer Time-Sharing—Everyman at the Console," *Fortune*, August 1967, pp. 88–91, 187–88, 190.

19. Fano, "Project MAC," p. 346.

20. Author's interview with Robert W. Taylor in Woodside, California, 8 August 1996.

21. Katie Hafner and Matthew Lyon, *Where Wizards Stay Up Late: The Origins of the Internet* (New York: Simon & Schuster, 1996), pp. 68–69.

22. Arthur L. Norberg and Judy E. O'Neill, *Transforming Computer Technology: Information Processing for the Pentagon, 1962–1986*, Johns Hopkins Studies in the History of Technology (Baltimore: Johns Hopkins University Press, 1996), pp. 16ff.

23. Roberts's quotations in this section are from the interview with Lawrence G. Roberts conducted by Arthur L. Norberg, 4 April 1989, Charles Babbage Institute.

24. Bolt Beranek and Newman, "Completion Report" (Cambridge, Mass., January 1978), p. III-14.

25. McKenzie and Walden, "ARPANET," p. 344.

26. On the general subject of the interconnection of subsystems through gateways, see Paul A. David and Julie Ann Bunn, "The Economics of Gateway Technologies and Network Evolution: Lessons from Electricity Supply History," *Information Economics and Policy* 3 (1988): 165–202.

27. Lawrence G. Roberts and Barry D. Wessler, "Computer Network Development to Achieve Resource Sharing," *Proceedings Spring Joint Computer Conference* (American Federation of Information Processing Societies, 1970), p. 543.

28. Paul Baran, *Reliable Digital Communication Systems Using Unreliable Network Repeater Nodes,* Report P-1995 (RAND Corporation, 1960), p. 1.

29. Baran quotations in this section are from the interview with Paul Baran conducted by Judy O'Neill, 5 March 1990, Charles Babbage Institute.

30. Communication from Kleinrock to the author, 30 December 1997. Kleinrock's dissertation was published as *Communication Networks* (New York: McGraw-Hill, 1962).

31. Norberg and O'Neill, *Transforming Computer Technology,* pp. 159–62.

32. Janet Abbate, "From ARPANET to INTERNET: A History of ARPA-Sponsored Computer Networks, 1966–1988," (Ph.D. diss., University of Pennsylvania, 1994), p. 44. I am indebted to Janet Abbate for several research essays about the ARPANET that she prepared as a graduate student assistant on this book project.

33. By means of the RFP, Roberts "went after the particular technology [he] . . . wanted," which led him to say later that the ARPANET "was my project in terms of design and control." Interview with Roberts conducted by Arthur Norberg, 4 April 1989.

34. BBN, "Completion Report," p. III-19.

35. Interview with Severo Ornstein conducted by Judy O'Neill, 6 March 1990, Charles Babbage Institute.

36. Interview with Robert Kahn conducted by Judy O'Neill, 24 April 1990, Charles Babbage Institute.

37. Interview with Frank Heart conducted by Judy O'Neill, 13 March 1990, Charles Babbage Institute.

38. Interview with David Walden conducted by Judy O'Neill, 6 February 1990, Charles Babbage Institute.

39. Interview with Frank Heart conducted by O'Neill.

40. Interview with Severo Ornstein conducted by O'Neill.

41. Interview with David Walden conducted by O'Neill.

42. Interview with Severo Ornstein conducted by O'Neill.

43. Interview with David Walden conducted by O'Neill.
44. Interview with Lawrence Roberts conducted by Arthur L. Norberg.
45. Interview with Severo Ornstein conducted by O'Neill.
46. Communication from Kleinrock to the author, 30 December 1997.
47. Interview with Severo Ornstein conducted by O'Neill, and communication from Frank Heart to the author, 2 December 1997. I am indebted to Heart for clarification on this and several other substantial points.
48. Interview with William Crowther conducted by Judy O'Neill, 12 March 1990, Charles Babbage Institute.
49. Interview with Leonard Kleinrock conducted by O'Neill.
50. Hafner and Lyon, *Where Wizards Stay Up Late,* p. 122.
51. I am indebted to Lawrence Roberts, president of ATM Systems, for this definition of "protocol" and related terminology.
52. Bryan Pfaffenberger, ed., *Que's Computer User's Dictionary* (Indianapolis: Que Corp., 1993), p. 135.
53. Interview with Vinton Cerf conducted by Judy O'Neill, 24 April 1990, Charles Babbage Institute.
54. Interview with Alexander McKenzie conducted by Judy O'Neill, 17 March 1990, Charles Babbage Institute.
55. I am indebted to Douglas McIlray for clarification of the concepts of signaling and layering.
56. Abbate, "From ARPANET to INTERNET," p. 59.
57. Interview with Leonard Kleinrock conducted by O'Neill.
58. Interview with Vinton Cerf conducted by O'Neill.
59. Janet Abbate, "Roles and Functions Within the ARPANET System," research essay, May 1991.
60. Hafner and Lyon, *Where Wizards Stay Up Late,* pp. 129–32.
61. Ibid., p. 157.
62. Ibid., pp. 175–76.
63. Interview with Robert Kahn conducted by O'Neill.
64. Interview with Alexander McKenzie conducted by O'Neill.
65. Hafner and Lyon, *Where Wizards Stay Up Late,* pp. 189, 194.
66. Janet Abbate, "Complexity and Control: Lessons from the Evolution of the ARPANET," research essay, 1992.
67. Summary data taken in part from Internet Timeline (1993) by Robert H Zakon, to be found at hobbes@hobbes.mitre.org. Janet Abbate has pointed out to me that other Department of Defense networks spun off from the ARPANET, among them two for the Defense Communications Agency and two for the National Security Agency.

68. Thomas P. Hughes, "Edison's Method," in *Technology at the Turning Point,* ed. William B. Pickett (San Francisco: San Francisco Press, 1977), pp, 5–22; Thomas Parke Hughes, *Elmer Sperry: Inventor and Engineer* (Baltimore: Johns Hopkins University Press, 1971); and David A. Hounshell, *From the American System to Mass Production, 1800–1932: The Development of Manufacturing Technology in the United States* (Baltimore: Johns Hopkins University Press, 1984).

69. Heart quotations in this section from interview with Frank Heart conducted by O'Neill.

70. McKenzie quotations in this section from interview with Alexander McKenzie conducted by O'Neill.

71. Author's interview with Robert Taylor in Woodstock, California, 8 August 1996.

VII. EPILOGUE: PRESIDING OVER CHANGE

1. The quotations are from Alfred Chandler, *The Visible Hand: The Managerial Revolution in American Business* (Cambridge, Mass.: Belknap Press, 1977), pp. 7–8

2. Harvey Brooks called my attention to the similarity of characterizing a black box only by its external interactions without regard to its internal works to the difference between thermodynamics and statistical mechanics.

3. For a subtle discussion of modernity, see Bruno Latour, *We Have Never Been Modern* (Cambridge, Mass.: Harvard University Press, 1993).

4. In setting up this chart, I have drawn on David Harvey, *The Condition of Postmodernity: An Enquiry into the Origins of Cultural Change* (Oxford: Basil Blackwell, 1989). On modern and postmodern characteristics, see also Robert Venturi, *Complexity and Contradiction in Architecture* (New York: Museum of Modern Art, 1966); Denise Scott Brown, "Between Three Schools," *Architectural Design* 60.1-2 (1990): 8–20; and Zygmunt Bauman, *Intimations of Postmodernity* (London: Routledge, 1992).

Index

Page numbers in *italics* refer to illustrations.

ablation, 127

academia, systems approach as subject in, 145, 149, 151, 152–54

Ackoff, Russell, 192–94

acoustics, 260

ADIS (Air Defense Integrated System), 42–43

ADSEC (Air Defense Systems Engineering Committee) (Valley Committee), 18–23, 24, 27, 61, 67

Advanced Research Projects Agency (ARPA):
Defense Department establishment of, 258
funding of, 258, 269
information network project of, *see* ARPANET
IPTO, 259, 263–64, 268–70, 276–77, 283, 297
micromanagement avoided by, 256–57
system-building focus of, 8

Advisory Committee on the Inner Belt, 202

Aerojet Engineering Company (Aerojet-General Corporation), 72, 83, 108, *110,* 186, 187–88

Aerospace Corporation, 138, 156

aerospace industry:
recession in, 170
sociotechnical problems studied by, 170, 185–89, 195
systems engineers in, 145

aircraft industry:
electronics development in, 114–15
military alliances with, 70–71, 73, 133
as prime contractor vs. associate, 89–91, 97–99, *100, 101,* 131–36

Air Defense Integrated System (ADIS), 42–43

air defense systems:
in Britain, 147–48
for long-range bombers, *see* Semiautomatic Ground Environment Project
OR approach to, 147–48
passive defense aspects of, 26, 28

Air Defense Systems Engineering Committee (ADSEC) (Valley Committee), 18–23, 24, 27, 61, 67

Air Force, U.S.:
Army separation from, 18, 24
base locations for, 157

Air Force, U.S. (*continued*):
  bureaucracy of, 103
  communication system developed for, 272–74
  ground-based defense system developed with, 17, 18, 23–24, 26
  institutional inertia within, 76–77
  operations analysis for, 150–51
  RAND established by, 154, 155, 159
  Science Advisory Board of, 18, 28, 67, 72, 81, 85, 92, 106
  weapons development and, 70–74, 75–76, 92–93, 155; *see also* Atlas Project
Air Force Ballistics Missiles Committee, 105
Air Material Command, 93
Air Research and Development Command (ARDC), 92–93, 137, 171
  WDD under, *see* Western Development Division
Albers, Joseph, 167
Allis-Chambers, 123
Altshuler, Alan, 202, 204
American Joint New Weapons Committee, 150
Anderson, Clinton P., 104–5
Antisubmarine Warfare Operations Research Group (ASWORG), 149
Antonelli, Kay McNulty, 55
Apollo lunar-landing program, 118
Applied Physics Laboratory, 145, 152–53
Archilochus, 120

ARDC (Air Research and Development Command), 92–93, 137, 171
  WDD under, *see* Western Development Division
Army Corps of Engineers, 227, 232
Army Operations Research Office, 152, 153
Army Ordnance Department, 73
Arnold, Henry (Hap), 71–73, 88, 94, 155
ARPA, *see* Advanced Research Projects Agency
ARPANET, 255–300, 285, 287
  computer science departments developed through, 256
  contractors on, 276–77, 279
  counterculture values expressed in, 13, 279, 303, 304
  e-mail on, 292–93
  former SAGE personnel involved in, 66–67
  host-to-host protocols of, 284–88
  IMPs created for, 276–83, 284
  informal organizational style of, 257, 278–79, 281, 286
  information highway initiated through, 5
  Licklider's groundwork for, 262–68
  management of, 294–300
  military needs and, 256, 257–58, 270, 279
  network of affiliated personnel in, 263–65, 270–71, 295
  1972 public demonstration of, 255, 291–92
  packet switching in, 272–75

postmodernity of, 303–4
Project MAC as precursor of,
265–67
research value of, 267, 271–72
telephone system as basis of,
257, 271, 274
testing of, 289–91
transdisciplinary teams utilized
in, 5
university sites of, 255, 264,
283, 286, 296–97, 298
artificial intelligence, 261
assembly-line components, 5
ASWORG (Antisubmarine Warfare Operations Research
Group), 149
AT&T, 90, 91, 112, 273–74
Atlas Project, 69–139
accuracy concerns in, 98,
131–32, 137
budget concerns on, 75–76,
103–6, 124
bureaucratic oversight
contained on, 103–7, 198
civilian advisory group on,
92, 96–99, 125, 126, 131,
151
concurrency implemented in,
107
contractors chosen for, 108–9,
110, 122–24, 127, 129–36
control room created for,
125–26, 245
duration of, 69–70
electronics and computers used
in design of, 91, 98, 110,
125–26, 128, 130
final assembly of, 108–9
guidance systems used in, 75,
98, 110, 135

managerial responsibilities on,
60, 64, 89–90, 96–102,
117–18, 124–26, 240,
256–57
military head of, 93–97,
99–109, 117–18, 124–25,
138, 198–99
missile deployment from, 136,
137
MX-774 transition to, 74
nose cone developed for, 75,
79, 110, 122–23, 124,
126–31
origins of, 70–84
propulsion system of, 76, 89,
108, 110
as prototype of systems
approach to large-scale projects, 70, 107–9, 118–19,
125–26, 138–39, 144–45,
237, 240, 245, 302
size of, 4, 70, 122, 125, 138,
257
Soviet threat as spur to, 79–82,
104–5
space launch vehicle based on,
137
systems engineering developed
on, 4, 58, 89–90, 101,
118–25, 257
Teapot Committee recommendations on, 84–92
testing procedures on, 76,
127–28, 131, 135–37
three products of, 70, 109
weight reduction in, 74–75, 89
Atomic Energy Commission, 80,
84, 85, 215
AUTODIN, 275
automation, 21

AVCO Manufacturing Corporation, *110*, 123, 124, 127, 128, 129, 133
Azusa guidance system, 135

Bacher, Robert, 128–29
Baldwin, Hanson, 79
ballistics, 74
  see also Atlas Project; intercontinental ballistic missiles
Baran, Paul, 272–75
Barker, Ben, 280
Barnhart, Ray, 210–11
Bartell, Bill, 280
batch process, 262
Bay Area Rapid Transit (BART), 213
BBN (Bolt Beranek and Newman), 260, 276–83, 284, 287, 288, 292, 296
Beard, Edmund, 77
Bechtel Corporation, 212, 213–15
Bechtel/Parsons Brinckerhoff:
  background of, 213–15
  CA/T project management by, 5, 212–13, 221, 231, 235–50
  EIS prepared by, 215–17
  as joint-venture firm, 212, 213
  management structure of, 236–37
  review process on, 251–54
  system building as focus of, 8
Beckman, Arnold, 113
Belaga, Julie, 217
Bell Telephone Laboratories, 25, 26, 27, 66, 86, 91, 98, *110*, 114
  on SAGE, 48–49, 52, 58, 62
Bendix Aviation, 86, 123, 124
Bendix Radio, 58

Benington, Herbert D., 57
Beranek, Leo L., 260
Berlin, Germany, highway construction in, 199
Berlin, Sir Isaiah, 120
Bernal, J. D., 146–47, 148
Bethe, Hans, 81
B-52 bomber, development program on, 138
Blackett, P. M. S., 146, 148–49
Bode, Hendrik, 86
Bolt, Richard, 260
Bolt Beranek and Newman (BBN), 260, 287, 292, 296
  IMP developed by, 276–83, 284, 288
BOMARC, 42, 61, 62, 65
bombers:
  missile development vs., 75, 81–82, 93
  OR study on, 155
Borg-Warner, 123
Bossart, Karel J. (Charlie), 74–76, 89
Boston, Mass.:
  Back Bay landfill of, 200
  CA/T plan in, see Central Artery/Tunnel Project
Boston Globe, 14, 223, 224, 225, 226, 229, 233, 251, 253
Boston Transportation Planning Review, 203
Bowles, Edward, 155
Boyd, Alan, 95
Bradbury, Norris E., 81, 92
Brett, George H., 94
Bridge Design Review Committee, 230–35
Brodie, Bernard, 156–57
Brookhaven National Laboratory, 149

Brown, Edmund G. (Pat), 170, 185
Brown, Gordon, 27
Brown, James P., 170
Brunetti, Jeff, 240–42, *241*
B-29 bomber, 155
Buderi, Robert, 32
budget operations, systems analysis on, 164–65
Bunker-Ramo, 276
Burckhardt, Jakob, 190
bureaucracy, inefficiencies of, 103–7, 113
Burroughs, *110*
Bush, Vannevar, *12*, 78–79, 150

Califano, Joseph A., *172–73*
California, systems approach to sociotechnical problems in, 170, 185–89, 195
California, University of, 153–54, 159, 264, 283, 286, 298, 299
California Institute of Technology, 18, 27, 71, 86, 110, 111–12
cargo cults, 169
Carnegie Institute of Technology, 145
Carnegie-Mellon University, 264, 298
Carol R. Johnson & Associates, 249
Case Institute of Technology, 154, 192
Center for Investigative Reporting, 215
Central Artery/Tunnel Project (CA/T), 197–254
    architectural design on, 216–17, 223, 225, 232–33, *236*, 246–50, *247*

Charles River crossing and, 199, 205, 218, 220, 223–35, *226, 228, 234, 236, 241*, 242
completion date for, 199, 235
contractors chosen on, 240, 241, 242, 243
control room operations on, 244–45
costs of, 199, 200, 207, 210, 220, 222–23, 235, 237–39, 241, 251, 253
design phases on, 239–44, *241*, 246–50
economic benefits of, 200, 207–8, 216
elevated central artery vs., 199, 200
environmental issues faced by, 14, 199, 200, 210–11, 215–23, 231, 232–33, 235, 241, 242, 304
federal funding of, 208–11, 224, 233–35
"green snake" replaced by, 199
harbor tunnel portion of, 199, 205, 207, 208
initiation of, 211–15
Inner Belt plan vs., 201–2, 203, 208
1995 management review on, 251–54
minority and female participation in, 14, 249
as open system, 197
origins of, 204–8
postmodernity of, 303, 304
project management on, 211–13, 214, 236–46, 251–54

Central Artery/Tunnel Project
(CA/T) (*continued*):
public participation in design
of, 213, 217–18, 220,
223–35
Scheme Z, 220, 223–30, *228,*
231–35
scope of, *198,* 199
social and political factors
included in, 13–14, 207,
209, 210, 216, 218–20, 221,
222, 227, 248, 304
time span for, 4, 199
traffic system in, 207, 227,
245–46
transdisciplinary teams on, 5
underground section of, *198,*
199, 204–5, 207, *243, 244,*
247
ventilation buildings in, 250
Century Freeway, 210
Cerf, Vinton, 286, 289, 293
C-5A transport, 163
change generation, computer
industry commitment to, 5
Chapman, Priscilla, 230
Charles River, bridge across, 199,
205, 218, 220, 223–35, *226,*
228, 234, 236, 241, 242
Chrysler, 124
Chu, Chuan, 38
civil government, systems
approach applied to, 141,
166–67
Clark, Wesley, 271, 275–76
Clauser, Francis, 73
Cold War:
defense research priorities in,
75, 81–82, 199, 255–56
end of, 107, 139
nuclear threat of, 192

technological efforts spurred by,
11, 16–17, 107
Collbohm, Franklin R., 92,
131–32, 155, 156
Collins, John F., 180–81
command, control, and communication (C$_3$), 257–58
Committee for Regional Transportation, 224–25, 231,
235
complexity:
cause-and-effect relationships
within, 182–83, 184–85
counterculture escape from,
189–91
of social problems, 143, 166–70
for systems analysis, 158
systems approach suited to,
142–43
Compton, Karl T., 17
computer technology:
in Atlas Project, *110,* 125–26,
128, 130
as enhancement of human
thought, 260–62
for information network, *see*
ARPANET
innovative problem-solving
approaches in, 5
magnetic-core memory in,
33–34, 35–40, *37, 39*
project orientation of, 7
as public utility, 255, 267
in real-time military information processing, 4, 16, 17,
19–21, 22, 27–28, 54, 144,
257–58
of software development,
55–58, 258, 284–88
in systems-analysis work,
158

of time-sharing, 40, 66, 260,
262–63, 265–68, 272
university departments in, 256
Whirlwind development in,
30–40
Conant, James, 29
concurrent development, 107–9
Congress, U.S.:
ARPA funding and, 269
Boston highway project funded
by, 208–11, 224
ICBM program supported in,
104–5
urban programs enacted by, 168
consensus, 279, 296, 304
Conservative Law Foundation,
222, 223, 224, 229, 231
Consolidated Vultee Aircraft Corporation (Convair), 73,
74–76, 80, 89–92, 97–99,
100, 108–9, 110, 123, 131,
132–36, 137
Cosell, Bernie, 280
counterculture, 13, 189–91, 194,
279, 304
Crawford, Perry, Jr., 31
crime prevention and control, 186,
187
Crocker, Steve, 286
Crowther, William, 278, 279,
280, 280, 281, 290, 291
C₃ (command, control, and communication), 257–58
Cybernetics (Wiener), 21, 178

Davidson, Ward F., 150
Davies, Donald, 275
DBE (disadvantaged business
enterprises), 249
DEC (Digital Equipment Corpora-
tion), 36, 67

Defense Ballistic Missiles Committee, 105
Defense Department, U.S.:
military research assigned to
service branches by, 73–74
missile development overseen
by, 73–74, 103
operations research efforts at,
149
research arm of, see Advanced
Research Projects Agency;
ARPANET
systems approach cultivated at,
11, 141, 160–66
weapons systems evaluated for,
149, 156, 162–63, 164–65
DeLauer, Richard, 118
Deltamax, 35–36, 38
Demonstration Cities and Metropolitan Areas Redevelopment
Act (1966), 168
Department of Public Works
(DPW), 201, 212, 215–16,
217, 220, 221, 248
DeVillars, John, 218–20, 230
Digital Equipment Corporation
(DEC), 36, 67
disadvantaged business enterprises
(DBE), 249
distributed adaptive message
block system, 273
Dodd, Stephen, 56
Dole, Elizabeth, 210–11
Donnelly, Brian, 208–9
Douglas, Donald, 63, 72–73,
155
Douglas Aircraft Company, 72,
109, 131, 133, 155, 156
DPW (Department of Public
Works), 201, 212, 215–16,
217, 220, 221, 248

Dukakis, Michael, 204, 206, 207, 208, 222, 223, 229, 231
Duke, William, 118
Dunn, Louis, 86, 117, 130–31

Eckert, J. Presper, 38
Edison, Thomas, 3, 6, 278, 296
Einstein, Albert, 85
EIS (environmental impact statement), 204, 210–11, 213
Eisenhower, Dwight D., 41, 64–65, 80–84, 86, 88, 92, 96, 105, 136–37, 157, 199
electric power, regional systems for, 270, 271
electronic mail (e-mail), 292–93
Eliot, T. S., 167
Ellsberg, Daniel, 162
Ellul, Jacques, 191, 194
English Channel tunnel, 214, 251
ENIAC, 55
Enthoven, Alain, 11, 159, 162–66, 191–92
environmental impact statements (EIS), 204, 210–11, 213
environmental issues:
    for construction projects, 14, 200, 204, 210–11, 215–23, 231, 232–33, 235, 241, 242, 304
    military projects and, 10, 215
    mitigations on, 221–23, 235
    systems approach on, 170, 186
Epstein, Elizabeth, 224
equilibrium economics, 158
Erie Canal, 15, 66
Everett, Robert R., 31, 32, 36, 48, 48, 49, 56, 59–62, 65, 67

Falcon missile, 113
Fano, Robert, 266, 267, 270

Federal Highway Administration (FHWA), 207, 209, 210, 211, 212, 220, 232, 235, 237, 250
feedback processes, 34, 177–79, 182
Ferenz, Ramona, 56
Finletter, Thomas, 40–42
fire-control devices, 178
Flagle, Charles, 153
Fletcher, James, 118
Floyd, Peter, 248
Flynn, William S., 239, 251, 253
Ford, Henry, 6, 159, 160, 296, 302
Ford, Henry, II, 160, 161
Ford Foundation, 155, 159
Ford Motor Company, 3, 123
Forrester, Jay, 7–8
    career of, 34–35, 59
    magnetic-core memory developed by, 33–40, 37
    on Michigan system vs. Lincoln plan, 45
    military use of computer technology proposed by, 17
    SAGE computer developed by, 48–52, 67
    on SAGE project management, 58–60
    on system dynamics, 141, 176–85
    urban model developed by, 171, 178, 180–83, 180
    in Whirlwind computer development, 28, 30–40, 48, 48, 56
Frank, Howard, 290
Fuller, Buckminster, 248

Gaither, H. Rowan, 159
Galbraith, John Kenneth, 202

GALCIT (Guggenheim Aeronautics Laboratory), 71, 86
game theory, 157
Gardner, Trevor, 81, 83–85, 86, 88–93, 99–101, 103–6
Geisman, Jim, 280
General Electric (GE), 73, 83, *110*, 112–13, 114, 123, 124, 127, 129–31, 133, 179, 277
General Motors Corporation, 83, 123, 302, 304
geodesic domes, 248
Gifford, K. Dun, 224–25, 235
Gillette, Hyde, 105
Glasser, Otto, 123–24
Glenn L. Martin Company, 73, 108, *110*, 137
Gödel, Kurt, 167
Goldberg, Richard, 224–25, 229
Goldwater, Barry, 167
Gordon, Cecil, 148
Great Society Programs, 141, 167, 173, 194, 304
Gross, Robert E., 114–15
Groves, Leslie, 101, 107–8
Guggenheim Aeronautics Laboratory (GALCIT), 71, 86
gunfire-control devices, 178

Hafner, Katie, 290
Haldane, J. B. S., 147
Halligan, Clair W. (Hap), 64
Harvard Business School, 183, 251
Harvard University, 27, 29, 277
Hayes, Dan, 202
Hayes, Munro, 38
Hay Market Pushcart Association, 217–18
Heart, Frank, 278–79, 280, 288–89, 291, 296, 297–98
heat sink, 127, 129

Hewlett, William R., 159
highway construction:
    federal aid policy on, 203, 204, 208–11
    presidential vetoes of, 209, 211
    public attitudes toward, 14, 227–28, 304
    as special-interest projects, 209–10
    see also Central Artery/Tunnel Project
Hill, Albert G., 27, 30, 46
Hitch, Charles, 156–57, 158, 161–62, 164, 165, 166
Honeywell, 277, 282
Hoos, Ida, 194–95
Hoover Dam, 6, 213–14
hospital management, 153
Housing and Urban Development (HUD), U.S. Department of, 168, 174, 175
Howard, James, 209
Howe, Mark de Wolfe, 202
Howe, Peter, 224, 225, 228, 229
Huddleston, Edwin E., Jr., 159
Hughes, Howard, 113, 114, 116
Hughes Aircraft Company, 21, 84–85, 86, 89, 90, 113–16, 118
Humphrey, Hubert, 143
Hyland, Lawrence, 86, 99

IBM (International Business Machines) Corporation, 38, 48–52, 54
ICBMs, see intercontinental ballistic missiles
ICBM Scientific Advisory Committee, 92
IMPs (interface message processors), 276–83, 280, 284, 289, 290, 291

industry:
  operations research applied to,
    148–49, 151
  system dynamics in, 179–80
information flows, 179, 182
information networks, *see*
  ARPANET
Information Processing
  Techniques Office (IPTO),
  259, 263–64, 268–70,
  276–77, 283, 297
Inner Belt project, 201–2, 203,
  208
intercontinental ballistic missiles
  (ICBMs):
  initial U.S. research on, 70–76
  U.S. defense efforts on, 15, 24,
    26
  see also Atlas Project; Minute-
    man missile; Titan
interface message processors
  (IMPs), 276–83, 280, 284,
  289, 290, 291
intermediate-range ballistic missiles (IRBMs), 109
International Business Machines
  (IBM) Corporation, 38,
  48–52, 54
International Conference on Computer Communication, 291
Internet, ARPANET as predecessor of, 67, 293–94
interstate highways, federal funding regulations on, 204
Ionata, Ed, 220
IPTO (Information Processing
  Techniques Office), 259,
  263–64, 268–70, 276–77,
  283, 297
Irvine, William, 95

Jackson, Henry, 104–5, 165
Jacobi Systems, 276
Jacobs, John, 64, 66
jet propulsion, 72
Jet Propulsion Laboratory, 72, 86,
  91, 117, 128
John, Richard, 205
Johns Hopkins University, 145,
  152–53
Johnson, Lyndon, 13, 79–80, 141,
  160, 167, 168–70, 172, 189,
  192, 194, 304
Joint Venture, *see* Bechtel/Parsons
  Brinckerhoff

Kahn, Robert, 277, 278, 279–81,
  280, 284, 289–93
Kaysen, Carl, 26, 183
Keller, K. T., 80, 102
Kelly, Mervin J., 25, 44
Kennedy, John F., 86, 160, 161,
  189, 192
Kerasiotes, James T., 233–35,
  253
Kidder, Tracy, 296
Killian, James, 23, 26, 28, 40–42,
  62–63, 64–65, 67, 88
Kimball, George, 151
King, Dan, 224
King, Ed, 208
Kistiakowsky, George B., 86,
  99
Kleinrock, Leonard, 256, 263,
  275, 281, 283–84, 283,
  289–90, 291, 298
Kossiakoff, Alexander, 145

Labour Party, 148
Lancellotti, Tony, 242–43, 242
Land, Edwin, 27

Lauritsen, Charles C., 27, 86, 125
Lawrence Livermore Laboratory, 80, 81, 82
Leach, W. Barton, 150
Lemley & Associates, 251
Lenoir, Timothy, 8
Licklider, Joseph Carl Robnett, 7–8, 27, 259–66, 259, 268–69, 276, 297, 298
Lincoln Laboratory, 28–30, 29, 40–46, 52, 55–56, 58–62, 64, 67, 91, 144, 145, 260, 269–70, 278
Lindbergh, Charles, 92, 99
Lindsay, John, 187
local governments, systems approach for, 170, 171, 194–95
Lockheed, 128, 133, 172, 186, 187–88, 195
Logan International Airport, 198, 199, 207, 208, 217, 221–22, 240
Loomis, F. Wheeler, 26, 29, 30, 40, 41
Lorsch, Jay, 251
Los Alamos Laboratory, 81
Lyon, Matthew, 290

McCarthy, John, 255, 266–67
McCone, John, 215
McCormack, James, 63, 64, 93
McDermott, John, 191
McKenzie, Alexander, 287, 292, 296, 299–300
McMillan, Edwin M., 159
McNamara, Robert, 11, 141, 160–65, 168–69, 191–92
McNarney, Joseph, 135

McNichol, Daniel, 205
magnetic-core memory, 33–40, 37, 39
magnetron tube, 149
Malina, Frank, 72
management:
    bureaucratic impediments to, 103–7
    by collective structure of military-industrial-university complex, 4, 5
    hierarchical, 5, 302–3
    prime contractor plan vs. systems approach to, 97–102, 100
    project orientation of, 7
    research and development vs., 5
    roles for engineers and scientists in, 4, 102
    systems engineering mode of, see systems engineering
    transdisciplinary teams in, 5, 145, 147, 156, 157, 158–59
"Man-Computer Symbiosis" (Licklider), 260–63
Manhattan Project, 6, 17, 25, 82, 84, 91, 101, 107–8
Marchetti, John, 32
Marcuse, Herbert, 191
Marill, Tom, 268
Massachusetts:
    Bay Transportation Authority, 212
    Department of Public Works, 201, 212, 215–16, 217, 220, 221, 248
    governors of, 204, 206, 222
Massachusetts Highway Department, 201, 232, 237–39, 241, 245, 251–52

Massachusetts Institute of Technology (MIT):
ARPANET links with, 263, 264, 266–67, 270, 277, 278, 286, 298
Cal Tech education vs., 111
Lincoln Laboratory, 28–30, 29, 40–46, 52, 55–56, 58–62, 64, 67, 91, 144, 145, 260, 269–70, 278
presidents of, 23, 86, 160
Radiation Laboratory, 6, 17, 25, 26, 28, 33, 112–13
rationale for military research at, 29
Sloan School of Management, 59, 176–77, 179
as system builder for SAGE, 15, 17, 23–30, 40–47, 62, 91
systems engineering programs at, 145, 153–54
Whirlwind Project at, 22, 27, 28, 30–46, 45, 46, 48, 48, 51, 56, 67, 118, 257–58, 284
Matador missile, 133
Maud Committee Report, 149
MEDINET, 277
Melanesia, cargo cults of, 169
Meltzer, Marlyn Wescoff, 55
Menn, Christian, 233
Mettler, Ruben, 118
Michigan, University of, 40–47, 154, 161
microwave research, 112–13
military:
command and control techniques developed by, 257–58
nonmilitary culture influenced by, 141–42

OR efforts for, 147–48, 149–51, 153–55
see also specific military services
military-industrial-university complex:
Air Force involvement of, 70–71, 73, 99–102
on Atlas Project, 99, 101–2
Cold War as spur for, 11–12
collective management by, 4, 5
competitive issues of, 62, 99
decision-making responsibility within, 99–101
history of, 9–13
ideological cohesion of, 11–13, 29
national culture shaped by developments from, 142
political tensions within, 88
RAND board drawn from, 159–60
Miller, Stanley, 230, 232
Millikan, Clark B., 86
Millikan, Robert, 71, 112
Minnesota Mining, 123
Minuteman missile, 137
missile research, 71–77
see also Atlas Project; intercontinental ballistic missiles
MIT, see Massachusetts Institute of Technology
mitigations, 221–23, 235
MITRE Corporation, 8, 62–64, 65, 66, 67, 144, 156, 257, 274
Morison, Elting, 77–78
Morrison, Herbert, 148
Morse, Philip M., 149, 151–52, 153
Moses, Joel, 3
Moynihan, Daniel Patrick, 189–90, 202, 227–28

Mueller, George, 118
Multics Project, 267
Mumford, Lewis, 191
Murray, Hubert, 248–49
MX-774 missile project, 74–76

NAA Aerophysics, 123, 124
National Academy of Sciences, 73,
    152
National Advisory Committee for
    Aeronautics, 113
National Aeronautics and Space
    Administration (NASA), 8,
    118
National Environmental Policy
    Act (1969), 204
National League of Cities, 167,
    180
National Physical Laboratory, 275
National Research Council, 152
National Science Foundation, 294
National Security Council, 77,
    105
NATO (North Atlantic Treaty
    Organization), 11, 192
Navaho missile, 76, 89, 133
Naval Ordnance Laboratory, 128
Navy, U.S.:
    in missile development, 73, 109
    OR efforts for, 149–50, 151,
        153
    Special Projects Office, 144
Needham, Joseph, 147
Network Measurement Center,
    289–90, 296, 298, 299
Network Working Group
    (NWG), 284–88, 292
Newman, Robert B., 260
New York State Urban Develop-
    ment Corporation, 174
New York Times, 11–12, 79, 93

Nike ground-to-air missiles, 62
Nixon, Richard, 13, 189
North American Aviation, 73, 76,
    108, 110, 123, 124, 133,
    186–88
North Atlantic Treaty Organiza-
    tion (NATO), 11, 192
Northrop Aircraft, 73, 76, 172
Northwestern University, 154
NSFNET, 294
nuclear weapons:
    deterrence strategy on, 157
    development of, 108, 149
    ICBM delivery of, 24, 26
    ideological opposition to, 88
    parallel development of, 108
    weight of, 79, 80–81, 82, 89,
        92
NWG (Network Working
    Group), 284–88, 292

Office of Naval Research (ONR),
    31–32
Office of Science Research and
    Development, 78, 83, 90
Oliver Aircraft, 123
Olsen, Kenneth, 36
O'Neill, Thomas P. (Tip), Jr.,
    208–10
open systems, 197, 203
Operation Castle, 92
operations research (OR), 9,
    146–54
    as academic study, 149, 151,
        152–54
    British origins of, 146–49
    defined, 142, 146, 154
    in hospital management, 153
    in industry, 148–49, 151–52
    military use of, 147–48,
        149–51, 153, 155

operations research (OR)
(*continued*):
reaction against, 192–94
system dynamics vs., 176–77
systems analysis vs., 154, 157,
164
transdisciplinarity used in, 147
Operations Research Society of
America, 152, 156
Oppenheimer, J. Robert, 85, 88,
101
optimization, 121–22
OR, *see* operations research
Ornstein, Severo, 278, 279–80,
280, 281, 282, 290–91

Packard, David, 113
Palmer, Thomas C., Jr., 251
Papian, Bill, 36
parallel development, 108–9
Parsons, William Barclay, 213
Parsons Brinckerhoff Quade &
Douglas, 212, 213
Partridge, E. E., 42–44
Pennsylvania, University of, 145,
153–54, 192–93
Perkins, James A., 160
Perry, Robert, 77
Peters, Tom, 191
Peterson Consulting, 251
planning, programming, and bud-
geting system (PPBS), 162,
164
Plesset, Ernst, 156–57
Polaris, 109
population growth, 183
postmodern projects, 303–5
President's Science Advisory Com-
mittee, 13
Project Charles, 23–28, 23, 54, 259
Project Hartwell, 25–26

Project Lincoln, 28–29, 40–46
see also Lincoln Laboratory
Project MAC, 66, 265–67, 278
projects:
in history, 6
interdisciplinary systems
approach to, 115
modern vs. postmodern, 303–5
open vs. closed, 197, 203
public responsibility concerns
on, 13–14
psychoacoustics, 259, 260
Puckett, Allen E., 86
Pugh, Emerson, 51–52
pulsed-circuit video techniques, 33

quantitative analysis, 157
Quarles, Donald, 84, 101

Radiation Laboratory:
alumni of, 25, 26
microwave research at, 112–13
systems approach used at, 26
wartime radar systems developed by, 6, 17, 28, 33
Radio Corporation of America
(RCA), 38, 48–49, 62, 114
Rajchman, Jan, 38
Ramo, Simon, 7–8, 87, 109
on anti-military ideological
shift, 12
Atlas management and, 98,
105, 106–7, 108, 117, 122,
124, 125, 129, 132–35, 138
background of, 84–85, 89,
110–16
colleagues evaluated by, 83,
106, 118
later career of, 138
managerial style of, 84–85,
108, 110, 113–16, 117, 135

on Soviet missile capability, 82
on systems engineering, 69, 89,
115, 119–20, 138
on Teapot advisory committee,
86–88, 90, 92
on urban problems, 168, 170
Ramo-Wooldridge Corporation:
Atlas managerial work given to,
64, 92, 97–103, 109,
117–18, 124, 130, 131–35,
137–38, 257
establishment of, 86, 116–18,
144
system-building focus of, 8, 119
Teapot Committee supported
by, 86–88
transdisciplinary approach of,
145
TRW absorption of, 138, 186
Rand, Ayn, 249
RAND Corporation, see Research
and Development Corpora-
tion
Raytheon Manufacturing Company, 48, 50, 276
RCA (Radio Corporation of America), 38, 48–49, 62, 114
Reagan, Ronald, 209–10, 211,
214
Redmond, Kent, 43–44
Rees, Mina R., 32
regional planning, 171, 174
Reich, Charles, 191
Remington Rand, 48, 49, 50, 110
research and development, committees utilized in, 18
Research and Development
(RAND) Corporation, 63
board of trustees for, 159–60
communication system
researched by, 272–74

Defense Department personnel
from, 161–66
establishment of, 73, 144, 154,
155–56
on long-range missiles, 80, 131,
157, 162
OR work of, 154
presidents of, 92, 131
research choices at, 156, 272
SAGE software design by, 56,
73
systems analysis developed at,
154–60
systems approach cultivated by,
11, 56, 73
transdisciplinary work of, 156,
158–59
urban problems studied by,
187
work atmosphere at, 156, 159
Rickover, Hyman, 107–8, 132
Ridenour, Louis, 23, 28, 67
Riggsbee, Jerry, 237
Rising, Hawley, 280
River Rouge auto production
plant, 6
Roberts, Lawrence, 263, 268 72,
275, 276, 283–91, 296, 297,
299
Rockefeller, Nelson, 174
Roosevelt, Theodore, 78
Roszak, Theodore, 190–91
Rowen, Henry, 169
Roy, Robert, 153
Royal Air Force, 147, 148, 150

SAB (Science Advisory Board), 18,
28, 67, 72, 81, 85, 92, 106
SAGE, see Semiautomatic Ground
Environment Project
Sage, N. M. (Nat), 17

Salvucci, Frederick, 7–8
    background of, 198, 201–2,
        203, 204–6, 221–22
    Charles River crossing and,
        223–24, 226–27, 228–30,
        231–32
    on environmental issues, 199,
        221, 227, 231
    federal funding and, 208, 211,
        224
    management choice of, 211–12,
        214
    political skill of, 198–99, 202,
        206–8, 210, 211, 221, 228,
        229–30
Samuelson, Paul, 26
Sandia Laboratories, 84
Sanford, Terry, 211
Sargent, Frank, 202–3, 204, 230
Schriever, Bernard, 7–8
    background of, 93–96, 137,
        198, 239
    on colleagues, 84, 86
    contractor relationships and,
        122, 131–35
    later career of, 137, 171–72
    managerial style of, 69, 102–9,
        110, 117–18, 122, 124–26,
        137, 138, 198–99, 237,
        299
    physical appearance of, 93, 95,
        96, 109
    SAGE systems engineer reports
        to, 64
    systems approach chosen by,
        96–102, 100, 119, 131, 133
    urban problems addressed by,
        171–76, 188, 239
Schriever, Dora Brett, 94
Schwiebert, Ernest, 78

science:
    government recognition of, 88
    social impact of, 147
Science Advisory Board, 18, 28,
        67, 72, 81, 85, 92, 106
scientific management, 8, 142
SDC (System Development Corpo-
        ration), 56–57, 63, 67, 144,
        188, 258
Semiautomatic Ground Environ-
        ment (SAGE) Project, 15–67
    architecture design of, 52–54, 53
    budget of, 17, 41, 56
    computer information process-
        ing implemented in, 4, 16,
        19–21, 22, 27–28, 30–40,
        48–52, 54, 70, 144, 257–58
    engineers in management roles
        on, 4
    as failure, 15, 64–66
    implementation of, 65
    initial conception of, 16–28, 20
    MIT as system builder for, 15,
        17, 23–30, 40–47, 62, 91
    naming of, 28
    radar function in, 22, 54
    software programs in, 55–58,
        73
    systems approach used in,
        21–22, 57–64
    as technology transfer agent,
        66–67, 70, 144–45, 302
    transdisciplinary approach used
        on, 15, 18, 145, 159
    university/military cooperation
        on, 15, 18, 23–27
Servomechanisms Laboratory, 17,
        22, 31
Shannon, Claude, 281
Shepard, David A., 160

Sherman, Forrest P., 25
shock tubes, 128
Shriver, Sargent, 168–69
Shultz, George, 214, 215
Sierra Club, 217, 225, 231
Silber, John, 217
Simms, William S., 78
Sloan School of Management, 59,
    176–77, 179
Smith, Thomas, 43–44
Snark missile, 76, 89, 133
social problems, systems approach
    to, 13, 143, 166–68, 170,
    171–76, 180–89, 180
Social Science Research Council,
    152
Soviet Union, see Union of Soviet
    Socialist Republics
Space-General Corporation, 186,
    187
space programs, 143
Special Projects Office, 145
Spence, Frances Bilas, 55
Sperry, Elmer, 34–35, 296
Sprague, Robert C., 44
Stanford Research Institute (SRI),
    264, 205, 206, 298, 299
Stever, H. Guyford, 27
Stokes, Carl B., 189
Strategic Missiles Evaluation
    (Teapot) Committee, 81–82,
    84, 85, 86–92, 93, 100
Stratton, Julius, 23, 24, 62–63,
    160
Strauss, Lewis, 85
Stull & Lee, 249
summer studies, 24–26, 203,
    259–60
Sutherland, Ivan, 263, 268, 269,
    283, 297

System Development Corporation
    (SDC), 56–57, 63, 67, 144,
    188, 258
system dynamics, 176–85
System Dynamics (Forrester), 141
systems analysis, 9, 154–66
    Defense Department use of,
        160–66
    defined, 142, 154
    development of, 154–60,
        163–64
    at RAND, 154–60
Systems Analysis Office, 163,
    165–66, 192
systems approach:
    criticisms of, 192–95
    defined, 21–22
    disillusionment with, 188,
        189–91
    key organizations in, 73,
        154–60
    1960s spread of, 141–95
    to social and civic concerns, 13,
        166–76, 180–89
    three varieties of, 142, 154
systems engineering:
    as academic subject, 145
    on Atlas Project, 4, 58, 70,
        89–90, 101, 107–9, 118–26,
        257
    CA/T implementation of, 5
    defined, 69, 142, 154
    development of, 4, 8–9,
        118–25, 143–45
    interdisciplinary problem solv-
        ing in, 120
    for large-scale projects, 70,
        107–9, 118–19, 125–26,
        138–39
    optimization in, 121–22

systems engineering (*continued*):
organizational issues and technical focus handled by, 7, 142
organizations focused on, 8, 144
in post–Cold War era, 138–39
primary functions of, 61, 142
scheduling process in, 118–19, 120–21
technical direction integrated with, 109, 122, *122*

Talbott, Harold, 83, 106–7, 116
Task Force on Antipoverty Programs, 168–69
Taylor, Frederick W., 8, 9, 302
Taylor, Richard, 230, 231
Taylor, Robert W., 268, 269–70, 272, 297, 298–99
TCP/IP (Transmission Control Protocol/Internet Protocol), 288, 293–94
Teapot (Strategic Missiles Evaluation) Committee, 81–92, 93, *100*
technology:
cultural reaction against, 190–91
future of government funding for, 256
institutional inertia as obstacle for, 76–79
postmodern, 303–5
U.S. prowess in, 3
Teilhard de Chardin, Pierre, 161
Teitelbaum, Ruth Lichterman, 55
Teller, Edward, 81, 85, 95
Tennessee Valley Authority, 6
Terhune, Charles H., 102, 117, 126
Thatch, Truett, 280
Thompson Products, 116

Thompson Ramo Wooldridge (TRW), 118, 138, 170–71, 186, 188
Thor missile, 109
Thorpe, Marty, 280
Time, 69, 87
time-sharing systems, 40, 66, 260, 262–63, 265–68, 272
Titan, 70, 108–9, *110,* 124, 131, 135, 137, 138
Tobin, James, 26
transdisciplinary teams, 5, 145, 147, 156, 157, 158–59
Transmission Control Protocol/Internet Protocol (TCP/IP), 288, 293–94
Transportation Act (1981), 209
transportation policies:
opposition to, 201, 227–28
systems analysis of, 186–87
Truman, Harry, 16–17, 18, 77, 80, 102
TRW (Thompson Ramo Wooldridge), 118, 138, 170–71, 186, 188
TX-2 computer, 283–84

Union of Soviet Socialist Republics (USSR):
missile technology of, 79–80, 81–82, 98, 104, 136–37
RAND as seen by, 156
scientists' ideological opposition to, 85
U.S. view of threat from, 11–12, 16–17, 255–56, 258
in World War II, 147
Urban, Inc. (Urban Systems Associates, Inc.), 172–75, 188
Urban Dynamics (Forrester), 181, 183

urban problems, systems
approaches to, 143, 167–68,
170, 171–76, 180–83, *180*,
185, 188–89, 201
U.S. Conference of Mayors, 167
USSR, *see* Union of Soviet Socialist
Republics
Utah, University of, 264, 283, 286

Valley, George E., Jr., 18–19, 22,
24–27, 30–32, 44–47, 52, 59
Valley committee (Air Defense
Systems Engineering Com-
mittee) (ADSEC), 18–23, 24,
27, 61, 67
Vandenberg, Hoyt S., 23
Venturi, Robert, 167
Viehe, F. W., 38
Vietnam War, 12–13, 167,
190–92, 303
von Kármán, Theodor, 18, 71–72,
83, 92, 95, 137
von Neumann, John, 27, 81,
85–86, 95, 157
ICBM advisory group chaired
by, 92, 96–99, 125, 126,
131, 134, 135, 157
V-2 missiles, 71, 74–75, 79

Waddington, C. H., 147
Wald, George, 194
Walden, David, 278, 279,
280–81, *280*, 283, 291
Wallace, Dave, 248
Wallace, Floyd Associates, 216,
246–50
Wang, An, 38
Warren, Dean S., 188–89
waste management, 186
Watson, Thomas J., Jr., 44
Watson-Watt, Robert, 147

weapons systems, systems analysis
applied to, 142, 162–63,
164–65
Weapons Systems Evaluation
Group, 149
Webster, William, 159
Weigle, Theodore G. (Tad),
236–37, 253
Weinberg, Robert, 228
Weinberger, Caspar, 214, 215
Weld, William, 222, 230
welfare system, systems analysis of,
186
Western Development Division
(WDD):
Atlas Project managed by, 8,
93, 109, *109*, 256–57
establishment of, 93
successor organization to, 137
Western Electric Company, 52,
58, 84
Westinghouse, 112, 124
Westway, 210, 227
Wheelon, Albert, 118
Where Wizards Stay Up Late
(Hafner and Lyon), 290–91
Whirlwind digital computer, 22,
27, 28, 30–44, *45*, 46, *46*,
48, *48*, 51, *56*, 67, 118,
257–58, 284
White, Kevin, 202
White, Robert, 9
Whiz Kids, 162, 164
Wiener, Norbert, 21, 178, 260
Wieser, C. Robert, 57
Wiesner, Jerome B., 27, 30,
31–32, 86
Wilson, Charles, 83, 107, 116
Wofford, John, 203, 230
Wohlstetter, Albert, 157, 158
Wolf, Alice, 230, 235

Wooldridge, Dean, 86–88, 87, 92, 114, 116, 178
World War II (1939–1945):
aircraft deployed in, 70–71
British air defense in, 147–48
operations research in, 147–48, 149, 151
technological research during, 6, 17, 95, 112–13, 153
World Wide Web, 294

Yarmolinsky, Adam, 169
York, Herbert, 81, 86, 92

Zacharias, Jerrold, 24, 25–26, 27, 29, 30, 92
Zachary, G. Pascal, 296
Zraket, Charles, 66
Zuckerman, Solly, 146–47, 148
Zuk, Peter, 237–39, 238, 251, 252, 253